The Mechanization of the Heart

Rochester Studies in Medical History

Senior Editor: Theodore M. Brown
Professor of History and Preventive Medicine
University of Rochester

The Mechanization of the Heart: Harvey and Descartes

Thomas Fuchs
Translated from the German by Marjorie Grene

 THE UNIVERSITY OF ROCHESTER PRESS

Copyright © 2001 Thomas Fuchs

All Rights Reserved. Except as permitted under current legislation, no part of this work may be photocopied, stored in a retrieval system, published, performed in public, adapted, broadcast, transmitted, recorded or reproduced in any form or by any means, without prior permission of the copyright owner.

First published 2001
by the University of Rochester Press

The University of Rochester Press is an imprint of Boydell & Brewer, Inc.
668 Mount Hope Avenue, Rochester, NY 14620, USA
and of Boydell & Brewer, Ltd.
P.O. Box 9, Woodbridge, Suffolk IP12 3DF, UK

Library of Congress Cataloging-in-Publication Data

Fuchs, Thomas, 1958-
    [Mechanisierung des Herzen. English]
    The mechanization of the heart : Harvey and Descartes / Thomas Fuchs ; translated from the German by Marjorie Grene.
        p.cm. -- (Rochester studies in medical history, ISSN 1526-2715 ; 1)
    Includes bibliographical references and indexes.
    ISBN 1-58046-077-1 (alk. paper)
    1. Cardiology--History--17th century. 2. Heart--History--17th century. 3. Blood--Circulation--History--17th century. 4. Descartes, Rene, 1596-1560. 5. Harvey, William, 1578-1657. I. Title. II. Series.

QP101.4 .F8313 2001
612.1'09'032--dc21

2001048046

British Library Cataloging-in-Publication Data
A catalogue record for this item is available from the British Library

© Suhrkamp Verlag Frankfurt am Main 1992

Designed and typeset by Christine Menendez
Printed in the United States of America
This publication is printed is on acid-free paper.

For my parents

Contents

Translator's Foreword ... *ix*

Author's Foreword ... *xvii*

Abbreviations ... *xviii*

A. HARVEY AND DESCARTES ... 1
   I. THEME AND BACKGROUND OF THE INVESTIGATION ... 1

   II. SUMMARY OF THE ARGUMENT ... 10

B. THE GALENIC PARADIGM AND ITS CRISIS ... 19
   I. FERNEL'S *SUMMA* OF GALENISM ... 21
      a) Corollary: The Doctrine of the Spirits ... 22

   II. THE CRISIS OF GALENISM ... 27

C. THE VITAL ASPECT OF THE CIRCULATION: WILLIAM HARVEY ... 33
   I. THE SCIENCE OF THE LIVING ... 35

   II. *DE MOTU CORDIS*: THE MOTION OF HEART AND BLOOD ... 43
      a) The Motion of the Blood ... 47
      b) The Motion of the Heart ... 48
         1) *Basis of the Motion* ... 48
         2) *Mechanism of the Motion* ... 51
      c) The Function of the Circulation ... 53
         1) *Blood and Heat* ... 53
         2) *Tendencies of Harvey's Physiology* ... 56
      d) Conclusion ... 59

   III. *DE MOTU LOCALI ANIMALIUM*: THE MOVEMENT OF LIVING THINGS ... 62
      a) *Calor* and *Spiritus* ... 64
      b) Muscles and Nerves ... 66
      c) *Sensus and Motus* in Harvey's Later Works ... 69
      d) Summary: Polarity and Movement ... 73

   IV. *DE GENERATIONE*: THE PRINCIPLES OF THE CIRCULATION ... 75
      a) The Early Stages of Ontogenesis ... 77
      b) The Regularities of Development ... 79

	c) The Blood as Principle	*83*
	d) *Circulatio* and *Generatio*	*86*
	e) Summary: Cycle and Polarity	*89*

D. THE MECHANICAL ASPECT OF THE CIRCULATION: DESCARTES
   AND HIS FOLLOWERS *115*
   I. CIRCULATION AND PHYSIOLOGY IN DESCARTES *115*
      a) Cartesian Science *116*
      b) The Body without Soul *122*
      c) The Physiological Mechanisms *126*
         1) *Motion of Blood and Heart* *126*
         2) *The Movement of the Spirits* *131*
      d) Conclusion *138*

   II. THE MOTION OF HEART AND BLOOD AFTER DESCARTES *141*
      a) Holland *146*
         1) *Henricus Regius* *146*
         2) *Cornelis van Hoghelande* *148*
         3) *Franciscus Sylvius* *150*
         4) *Theodor Craanen* *153*
         5) *Cornelis Bontekoe* *154*
         6) *Stephen Blancaard* *156*
      b) England *157*
         1) *Early Writers* *158*
         2) *Thomas Willis* *160*
         3) *Richard Lower* *163*
         4) *John Mayow* *165*
      c) Other Writers: The Latency of the Vitalistic Aspect *168*

E. VITALISM AND MECHANISM BETWEEN 1700 AND 1850 *197*

F. A LOOK AHEAD *225*

Bibliography *233*

Index of Names *243*

Translator's Foreword

History is always more complicated than it looks at first sight. In this book, Thomas Fuchs illustrates that thesis with remarkable clarity. Let me anticipate, with some commentary of my own, the major moves of the story he has to tell. As the title indicates, he is first comparing the views of Harvey and Descartes about the heart and blood, and then tracing the way those opposing views—both accepting the circulation, but differing on the motion of the heart—were received, revised, rejected, or renewed in succeeding generations by medical writers in various parts of Europe.

First then, against the background of the collapse of Galenism, Fuchs examines Harvey's approach to cardiac and circulatory physiology, not only through the text of the *De Motu Cordis*, but through a consideration of all his surviving works, especially the essays on generation and also the recently published, but not much discussed, treatise on the local motion of animals. As the discoverer of the circulation, Harvey is rightly celebrated as the founder of modern physiology. But is he a "modern" thinker? Yes, he does use quantitative arguments at one point in the *De Motu Cordis*. But, if he is rejecting Galenic doctrines, he is doing so through a return to Aristotle—and what could be less modern than that? But that "fact", too, is complicated. His teachers in Padua before him, and Harvey himself, were Aristotelians in that they knew well, admired, and to some extent accepted the insights offered by Aristotle's biological writings. Taking Aristotle—as, indeed, Georges Cuvier was to do two centuries later—as the founder of comparative anatomy, they were happy to follow in the footsteps of a great biologist. It should be noted, however, that that does not mean Harvey was an "Aristotelian" in the sense in which seventeenth century scholastics were followers of "the Philosopher." On the European continent, at least, university students had to be trained in the elements of Aristotelian philosophy before they proceeded to the higher discipline of medicine, law, or theology. They were taught through recent commentaries on the works of Aristotle, filtered through centuries of debate and revision—and usually not on the biological works. They had to learn to think, and write, in terms of substance, form, and matter as their basic concepts, relying on notions like substantial form, real qualities, and the like. Harvey's undergraduate training at Cambridge was probably less rigid, and his years at Padua brought him into contact with the best practitioners and theoreticians in contemporary medicine. What characterized this kind of Aristotelianism was what Fuchs calls its vitalism: its passion for the careful and intense study of the living as living; its sense of the uniqueness of all that is alive. "Vitalism," of course, is usually a term of contempt; unfortunately, I cannot think of a substitute. Again, what it involves is accuracy, devotion, industry in the study of the details of living structures and processes. It was Aristotle who, in our tradition, initiated this kind of study, and it is this practice that Harvey's teacher, Fabricius, and Harvey himself were seeking to carry forward. On the face of it, to modern physiologists reared in a mechanistic tradition, in which organisms are viewed as subject to the physico-chemical laws that govern all the phenomena accessible to science, such discourse sounds rather quaint, to say the least. For Aristotle, it was the heart that was uniquely the seat of life; for Harvey, it is the blood itself, which shows itself in the embryo as the first seat of life, even before the formation of the heart. Especially if we study the *De Generatione* alongside the *De Motu Cordis*, we see, in Fuchs's words, that "[t]his primary organic substance, the 'life-stuff' blood, finally appears as ensouled through a universal principle of nature, in which Harvey describes the *primum efficiens* of generation as well as of the circulatory process." What is a modern cardiologist to make of such pronouncements? Clearly, those who want to see Harvey as the first modern medical theorist have to admit that he is saddled with a lot of "pre-scientific" baggage that he unfortunately carried with him. Better, Fuchs argues, to try to understand Harvey's basic attitude to his profession—comparative anatomy as well as medicine or comparative anatomy in the service of medicine—in his own terms, in terms of the thought-style (a phrase introduced by Ludwig Fleck) that characterized his approach to his work.

Two further points need to be mentioned here, one general and the other a particular application of it. One lesson Harvey had learned in Padua was the importance of following "the way of the anatomists," placing inspection (*autopsia*—that is, seeing for oneself) ahead of doctrine. The latter may be nobler, but the former is more certain. As he put it in the *Second Discourse to Riolan*:

> ... This is what I have striven, by my observations and experiments, to illustrate and make known: I have not endeavoured from causes and probable principles to demonstrate my propositions, but, as of higher authority, to establish them by appeals to sense and experiment, after the manner of anatomists.[1]

Even though, in the introduction to the *De Generatione*, he refers with respect to Aristotle's account of scientific method in the *Posterior Analytics*, where we advance from sense perception to indubitable first principles and deduce our explanatory conclusions from them, it is the starting point and the ultimate return to concrete, exact observation that he finds most significant.

Secondly, in particular, it was through vivisection in a great number of differing organisms that he had established his conclusions about cardiac (and arterial) motions and about the pulmonary transit. And, in the eyes of his contemporaries, these conclusions were truly revolutionary. It was generally agreed that when the heart hardens and strikes the chest, it is in diastole, and its limp phase is its systole. "Diastole," after all, means stretching. If you pick up a piece of string, for example, it hangs there limply, but if you stretch it out to wrap a package, it becomes taut and hard. Obviously, that is what happens to the heart when it beats. Remember, muscular contraction was not well understood at this time and involuntary muscle even less well. It would take some time and controversy to establish the heart's status as a muscle—a muscle that beats from birth to death and, in an excised heart, even after death. Absurd! But never mind those conceptual difficulties. Harvey had looked at, had examined carefully, the beating hearts of a great number of animals and had come to the firm conclusion that "...the very opposite of the opinions commonly received, appears to be true...the motion which is generally regarded as the diastole of the heart, is in truth systole."[2] It is this thesis, together with the pulmonary transit, that is established in the first seven chapters of the *De Motu Cordis;* the circulation appears only in Chapter VIII. There, Harvey asks: if the blood is expelled from the heart in a "forceful systole," where does all the blood go? Perhaps in a circle. As becomes apparent in the case of Descartes and his successors, the circle is easier to accept than the reversal of diastole and systole.[3]

After his thorough and stimulating presentation of Harvey, with his vitalistic bent, Dr. Fuchs proceeds to an examination of the aims and principles guiding Descartes's presentation of Harvey's views: his acceptance of the circulation and the disagreement with Harvey on cardiac motion. Although Descartes was not trained in anatomy, let alone medicine, he did do a number of dissections and some vivisections—including one that he said "cut the throat" of Harvey's theory of the heart beat. His overall aim, however, contrary to Harvey's, was "to lead the mind away from the senses," to separate anything ensouled or soul-like neatly from the physical, or physiological, and to establish a mathematical physics that would explain in terms of clear and distinct ideas all the operations of the world of bodies spread out in space—or better, perhaps, constituting space, since "extension, spread-out-ness" is just what Cartesian bodies are. And that includes living things, which, bereft of any special "vitality," are just so many spatial units intelligible solely—and wholly—through the laws of nature God has established and Descartes has discovered.

Now, in these terms, the circulation is appealing enough. Given the structure of the valves (which Fabricius had discovered, but misinterpreted), its makes good physical sense to have the blood go round and round. But that mysterious contraction at the moment of hardening? Wouldn't Harvey need some kind of "pulsative force" to make it happen? On the other hand, if the heart is a balloon-like furnace, stretching so that the (rarefied) blood overfills it and escapes into the arteries: in that case, we need only the good physical process of rarefaction to account for its beating. What causes the rarefaction? Well, all physicians know, says Descartes, that the heart is hotter than the rest of the body. (Here, again, Harvey was the heretic: he believed in vital heat, as most physicians did, but he located it in the blood as such rather than in the heart.) The heart contains, Descartes explained, a hidden fire like that in badly cured hay, a source of heat sufficient to cause instantaneous rarefaction, and so, the familiar stretching of the heart that is called diastole. That explains why, as many physicians also agreed, the blood changes color in the heart: it gets thinner. (Harvey did observe that the change of color occurred in the lungs, but alas, before Lavoisier, no one, despite active speculation, could make much sense of that.) Moreover, as Descartes reported in his correspondence with a Dutch physician, Plemp, or Plempius, if you cut the tip of a young rabbit's heart, you can see and feel the opening enlarging when the heart beats. In fact, that is the experiment that cuts the throat of Harvey's view. Indeed, Plemp, who had raised numerous objections to the notion of the circulation, was converted, presumably by Descartes's arguments (and observations!) and in a few years himself published a work on the circulation, borrowing, to Descartes's indignation, the Cartesian reading of the heart beat, and even distorting it in some respects.[4]

Something like this happened in a great many instances, which are reported here in illuminating detail. Of particular interest, for example, is the case of George Ent, one of Harvey's most ardent English disciples who, while praising the master for his discovery of the circulation, could not bring himself to accept Harvey's reversal of diastole and systole. If Descartes appealed to many through the simplicity and sweep of his mechanistic thought-style, hoping to bring all of nature under the governance of a set of universal laws, I believe his position proved more acceptable also because, in its medical details, it was more conservative. This point is not stressed by Dr. Fuchs, but it is, I believe, not incompatible with the story he is telling. A reigning thought-style often contains what Fleck call "pre-ideas" that will become more articulate when a new thought-style succeeds it. Surely, there are also "post-ideas" that linger in a new approach and make it easier to accept. That was the case, I believe, with Descartes's new mechanistic reading of cardiac physiology. He retained the "facts" commonly accepted by medical theorists: the heart is hotter than the rest of the body, the blood changes color there, and when the heart hardens and strikes the chest, it is stretching, that is, it is in diastole. In all these respects, Harvey wanted them to deny what everybody knew.

However that may be, it is certainly the case that it was frequently in Cartesian terms, not those of Harvey, that the circulation was accepted or, as Fleck would have put it, became not just a discovery, but a fact. I have mentioned the case of George Ent because his standing as Harvey's friend and disciple makes it especially striking, but Dr. Fuchs provides his readers with a report of such cases in each major European medical establishment. Even when some of the details of Descartes's account proved untenable, medical writers still tried to retain his fundamental approach. For example, in 1676, the Dutch physician Cornelis van Hoghelande declared, "...we are of the opinion that all bodies, however they behave, are to be viewed as machines and that their actions and effects...have to be explained only according to mechanical laws."[5] This is far, indeed, from Harvey's Aristotelian vitalism.

After his survey of the reception of the circulation up to 1700, Dr. Fuchs considers the period from 1700 to 1850, where he traces the interpenetration of mechanistic and vitalistic themes, with vitalism reigning for a while, to be replaced in the mid-nineteenth century by a largely mechanistic thought-style that still continues to dominate, though with latent vitalistic elements. Finally, he brings his history up to the present, adding a brief look ahead in conclusion. In these developments, it is ironic that when vitalistic themes reenter medical discourse, they are presented as replacing Harvey's more mechanical approach; the thought-style that had in fact characterized the great discoverer is largely forgotten, and he is still seen through the very Cartesian perspective that had distorted the original acceptance of the circulation.

I have referred several times to Dr. Fuchs's reliance on Fleck's concept of the "thought-style." To a professional philosopher like the present writer, his historiography is of special interest. He states it explicitly at the start of the book. Quoting a number of interpretations that try to assimilate Harvey to a linear series of advances, in which "the scientific method" is being triumphantly applied, he rejects such Whiggish readings. Thomas Kuhn's concept of the "paradigm" has largely replaced that linear view, and Fuchs does apply it to the case of Galenic tradition. In general, however, he finds Fleck's "thought-style" better applicable to his material. What interests him chiefly, as I have been indicating, is the complex interaction through several centuries of two thought-styles, vitalism and mechanism. The variants on each of these, of course, are manifold, and Dr. Fuchs has gathered information on an imposing array of them. (There need not be, he says, always just two such perspectives involved; that happens to be the case for the area he is reporting on.) Fleck's model seems to me an excellent one to follow. But in addition, Fuchs adduces a further consideration, which is, I believe, original to him. He suggests that it is not only different perspectives on the part of different workers that are involved, but also a complexity in the subject matter itself, which produces such diverse readings. There are aspects of a complex reality like that of the human (or should we say the mammalian?) circulatory system, which inevitably call forth one emphasis or another in specialists's approach to it. Such complexities, Fuchs argues, are never wholly exhausted by any special scientific approach, so that, however sophisticated our science becomes, a polarity like that of vitalism-mechanism, which has characterized the investigations of the past, is likely to continue in some way into the future as well. Some things are just more complex than any one theory can comprehend.

So, I may add in conclusion, is its history. Nobody, I suppose, can get it all quite right to everybody's satisfaction. Dr. Fuchs has been very generous in allowing me to suggest some amendments, especially to his accounts of Descartes. And I hope I may say without disrespect that I cannot really agree with him in his Pagelian reading of Harvey on circles. Walter Pagel found Harvey to be deeply influenced, not only by Aristotelian biology, but also by Aristotle's vision of the eternal circling of the celestial spheres. It is true that Harvey occasionally uses such rhetoric, referring to the heart, for instance, as the sun of the body, and so on. And it is true that when he does speak of the heavens, he speaks entirely traditionally, apparently in total ignorance of the new astronomy that was developing in his own day. But his answer to the question whether the blood evicted by the forceful systole may not be moving in a circle does not seem to me to need any such cosmic support. Indeed, Harvey rather dislikes astronomy than otherwise: if astronomers are to know anything about their subject, Harvey says mockingly, they will have to palpate the heavenly bodies and really see what they are talking about as anatomists do with their objects.[6] In other words, I don't know how much the great anatomist needed to look at the heavens to motivate his rendering of what he found here on earth. However, to say this is not for a moment to question Dr. Fuchs' interpretation of Harvey as a vitalist or his stress on rhythms and cycles that characterize whatever is alive. Indeed, the whole story told here provides fascinating and, for the most part, convincing reading both about the writer's chief protagonists and about the broad range of lesser figures who received and modified their heritage in so many different ways.

In translating *Die Mechanisierung des Herzens*, I was assisted most generously by its author, whom I consulted about medical technology and, as I have just admitted, chastised about what I thought to be errors in his account of some Cartesian texts. It was a pleasure to work with him. I also owe a great deal to my colleague, Dr. Mark Gifford, who read the whole manuscript critically and whose expert assistance was indispensable in the rendering of quotations from Aristotle. I am also grateful to the University of Rochester Press for including the book in their series on the history of medicine. It should offer, I believe, a rich and refreshing addition to our literature on the history of cardiac and circulatory physiology as well as to a developing body of work in a historically oriented philosophy of science.

Marjorie Grene
Blacksburg, Virginia
July, 2001

Notes to Translator's Foreword

1. William Harvey, *Second Discourse to Riolan* in *Works*, trl. R. Willis, p. 134 (See bibliography).
2. Willis, p. 22.
3. For an analysis of the structure of the *De Motu Cordis* and the probable order in which the chapters were written, see Robert G. Frank, Jr., *Harvey and the Oxford Physiologists*, Berkeley: University of California Press, 1980, p. 11, p. 37.
4. See Descartes, *Oeuvres*, Adam and Tannery edition, I, 399-401, 496-499; II, 62-69, 343-345.
5. See Chapter D II, n. 20.
6. Willis, p. 124.

Author's Foreword

The present study may be read and considered from different points of view: as a comparison of two authors both important for the history of medicine and biology, but with opposing conceptions of the world; as a contribution to the history of physiology from 1600 to 1850; as a study in the mechanization of the body in one of its central organs, namely the heart; as an application of the distinction between scientific *discovery* and *fact* as explained by Ludwig Fleck's theory of science on William Harvey's discovery of the circulation; and finally, as a case study on the conflict of different views, paradigms, or "thought-styles" over a longer period of the history of science.

Though I hope that the multitude of aspects to be followed will not only keep the study readable, but will even increase its attraction, a detailed introduction will try to convey the necessary overview on the topics and the course of the investigation. Moreover, summaries at the beginning of each chapter are meant to make the reader's orientation easier and to allow for a selection if necessary.

The present English edition of my book first published in Germany in 1992 would never have come to birth were it not for the insistence of Prof. Marjorie Grene, my translator and benevolent critic. I want to thank her cordially for the idea to translate the German text as well as for her energy and faith to realize this project herself, regardless of her advanced age. May she continue her work in the history and philosophy of science for a long time still, and may other authors profit from her scholarship as I did!

Thanks also go to my editor at the University of Rochester Press, Timothy J. Madigan, for his spontaneous acceptance of this project and his continuous support during its realization. I wish to share my pleasure in the publication of this book with all who have contributed to it in any way.

Thomas Fuchs
Heidelberg
July, 2001

ABBREVIATIONS

Aristotle is cited from Immanuel Bekker's *Aristotelis Opera* (Berlin, 1831-1870), English translations by Mark Gifford and Marjorie Grene.

Descartes is cited from the *Philosophical Writings of Descartes*, edited by John Cottingham, Robert Stoothoff, Dugald Murdoch, and Anthony Kenny. Cambridge, 1985-1991, 3 volumes (abbreviated CSMK), or from his *Oeuvres*, edited by C. Adam and P. Tannery. Paris, 1897–1910 (abbreviated AT).

Harvey is cited (with the exception of his *Praelectiones Anatomiae* and *De Motu Locali Animalium*) from: Harvei, Guglielmi. *Opera, Quibus Praefationem addidit Bernardus Sigfried Albinus*. Leyden, 1737. The second number gives the corresponding place in the translation by R. Willis: *The Works of William Harvey*. London, 1847 (abbreviated W).

AP	Aristotle. *Analytica Posteriora*
AT	Descartes. *Oeuvres*. Ed. Adam and Tannery
CSMK	Descartes. *Philosophical Writings of Descartes*
DA	Aristotle. *De Anima*
DCH	Descartes. *Description du Corp Humain*
De Iuv.	Aristotle. *De Iuventute*
De Resp.	Aristotle. *De Respiratione*
DG	Harvey. *De Generatione Animalium*
DGA	Aristotle. *De Generatione Animalium*
Disc.	Descartes. *Discours de la Méthode*
DMA	Aristotle. *De Motu Animalium*
DMC	Harvey. *De Motu Cordis*
DML	Harvey. *De Motu Locali Animalium*
DPA	Aristotle. *De Partibus Animalium*
ER	Harvey. *Exercitationes duae Anatomicae de Circulatione Sanguinis ad Johannem Riolanum filium*
M	Mayow. *De Motu Musculari*

Med.	Descartes. *Meditationes de Prima Philosophia*
PA	Descartes. *Les Passions de l'âme*
PP	Descartes. *Principia Philosophiae*
R	Mayow. *De Respiratione*
S	Mayow. *De Sal-nitro et Spiritu Nitro-aero*
TH	Descartes. *Traité de l'homme*
W	Harvey. *The Works of William Harvey*, translated by R. Willis

A

Harvey and Descartes

> Why may not the thoughts, opinions and manners now prevalent, many years hence return again after a long period of neglect?
>
> —**William Harvey,** *De Generatione Animalium.*[1]

I. Theme and Background of the Investigation

William Harvey (1568–1657) and René Descartes (1596–1650) can be equally well seen as pioneers of modern medicine.

At first, this statement may meet opposition. The great physiologist and discoverer of the circulation of the blood on the one hand, the self-taught medical amateur on the other—how could they be of comparable significance for the development of medicine? Harvey has always assumed a place of honor in the history of medicine, and modern physiology recognizes him as its forebear.[2] Descartes, on the other hand, is referred to chiefly as a precursor of the iatrophysicists, whose exaggerated and oversimplified mechanism is routinely greeted with a condescending smile.[3] In any event, Descartes has been granted recognition at most in the more restricted domain of neurology or physiology, say, for his doctrine of reflexes or of the passions. Otherwise, his physiology has usually been subject to the verdict already enunciated by Claude Bernard: "a phantasy physiology, almost entirely invented."[4]

Yet Harvey's position as the hero of modern medical science does suffer from a certain weakness, for, as his historiographers have to admit, his view of the world and of man was heavily influenced by Aristotle and even took a vitalistic form; and against all attempts to attribute this side of Harvey as far as possible to his late work, that is, to the *De Generatione Animalium* (1651), W. Pagel has shown how heavily even the discovery of the circulation rested on Aristotelian natural philosophy and methodology as well as on the speculative heritage of the Renaissance.[5] Thus there arises what at first sight appears to be a paradoxical consequence: one of the most important foundations of modern physiology is profoundly indebted to what are, from today's perspective, "pre-scientific" presuppositions and ideas.[6] However, it will be a primary goal of the following investigation to demonstrate with evidence drawn from his chief works *that not only Harvey's fundamental perspective, but also his interpretation of the process of circulation itself is of a thoroughly vitalistic nature, and thus deviates in essential features from today's conception.* In the course of this study, it will also become evident that Harvey was led by his discovery to a *new* form of vitalistic approach, which could no longer be absorbed within the traditional Aristotelian and Galenic framework, and which itself became the starting point of a tradition of thought culminating in the vitalism of the eighteenth century.

Descartes, on the other hand, not only sketched the first post-Galenic physiology in his *Traité de l'Homme* (*Treatise on Man*, 1632), but through his philosophical and scientific writings—as we will show more fully—influenced the medical view of man in far-reaching ways that are still being felt today. Only a Cartesian basis makes possible a physiology that is, in principle, independent of inner experience and is thus in the modern sense *scientific*.

Now, this modern physiological tradition begins at the precise point at which Harvey's vitalistic views arrived with the newly discovered circulation of the blood, in which Descartes was the first to recognize a decisively effective lever for carrying out a mechanical interpretation of the organism. Thus, at this central point, two opposing points of view overlapped, but one of them became so entirely dominant in the years that followed that it almost wholly concealed the other. Investigation of the sources will demonstrate that from the start, the reception of Harvey occurred essentially under Cartesian influence. This reception has shaped our picture of Harvey until today; even vitalistic physiologists later believed they must do battle against what was assumed to be Harvey's mechanistic interpretation of the circulation.[7] We can, therefore, formulate a further thesis: *Harvey's discovery, as well as Harvey himself, was and is seen chiefly from a perspective that was determined to a great extent not by him, but by Descartes.*

If, nevertheless, Descartes holds a comparatively lower rank in the history of medicine, this is to be attributed above all to the traditional vantage point of historians of science, which measured the significance of its protagonists, not by the development and acceptance of new viewpoints, but primarily by counting objectively specifiable discoveries. From the multifariously branching and interpenetrating development of the sciences, only those "hard" results were selected whose consequences could then appear as progress toward the present state of knowledge. Earlier, in part quite different, interpretations of the phenomena, on the contrary, counted as historically determined additions or curiosities or, at best, as surmountable presuppositions of particular "advances." For this kind of approach, Descartes has naturally little to offer since his significance lies not in such particular advances, but in the conception of a point of view which henceforth defined *what it was that was thought to count as an advance in the first place*, and which still frequently, if unconsciously, underlies the correspondingly selective view of medical historians. That is also the explanation of the picture of Harvey—at best one-sided—that prevails to this day.

In the meantime, in contrast to this tradition, and stimulated not least by T.S. Kuhn's *Structure of Scientific Revolutions* (1962), another conception of the history of science is beginning to take hold, which regards it as a discontinuous process, consisting of a sequence of "incommensurable" points of view or "paradigms." From this standpoint, the upheavals in this process are occasioned less by factors internal to science than by those that are external and, from an internalist point of view, "irrational"—namely, through the change of social and cultural structures, of life styles, attitudes, and world-views. It follows, from such a conception, that the point of view dominant today be no less recognized in its time-bound nature, compared with earlier views and, at least up to a point, relativized.

Kuhn's concept of "paradigm" is still closely oriented to the physical sciences and its central theories. In accordance with that orientation, it is relatively narrowly defined, as we shall see. As early as the 1930s, Ludwig Fleck (1896–1961) had developed a rather more inclusive conception; Kuhn himself builds on Fleck's doctrine of "thought-style" and "thought-collective."[8] By thought-style, Fleck understands a collective, gestalt-forming kind of perception and a specific system of concepts bound up with it, the parts of which cannot be transferred without change of meaning to a different thought-style. Only a comparative investigation of thought-styles, such as Fleck demands, is to allow us, from a more inclusive point of view, to bring verbally identical or cognate concepts from different thought-styles into relation to one another, to follow them in their development, and to look for common characters or differences in their intensional structures. According to Fleck, such an investigation often comes up against "pre-ideas," as germinal, still polyvalent concepts or ideas of a culture, the "motifs" of which can always be revived and developed further in divergent directions.[9] In this way, for example, Fleck sees in such ideas as "atom," "element," or "pathogenic agent," "developmental rudiments of modern theories."[10] Similarly, in the course of the present investigation, such "pre-ideas" will be found in the concepts, say, of "vital heat" (*calor innatus*), of the "spirits" (*spiritus*), or even in the concept of the "circulation" itself; and the different ways they were adopted or (re)interpreted from the perspectives of Harvey and Descartes will prove to be an essential moment in the emergence of the two viewpoints here being compared.

Closely related to the concept of "thought-style," and just as significant for us, is Fleck's distinction between "discovery" and "fact." In connection with the history of the Wassermann reaction as an indicator of syphilis, Fleck describes scientific discovery as a kind of anomaly: what the investigator perceives or recognizes is not immediately absorbed in the dominant thought-style; it does not fit into the general conceptual system.[11] One could also say that the discoverer still has to do with a singular phenomenon, which he confronts purely as an individual. But on its way to assimilation and recognition by the "thought-collective," the discovery is, at the same time, quietly transformed, adapted and "depersonalized."[12] This process "removes its [i.e. the discovery's] individual idiosyncrasy, shakes it loose from its originating, psychological, and historical contexts, and adapts it to the schemata of the total system;" while in this process, the system itself also undergoes a more or less perceptible modification.[13] Thus it is only in a *social process* that the scientific fact first arises, even if, with hindsight, it seems to contemporaries or to successors to have been already present as such at the moment of discovery.

This distinction is also fundamental to our investigation. As we shall see, Harvey's discovery of the circulation was not spared such "depersonalization" while being established in the intellectual context of its day. In any event, it belonged to the special class of discoveries whose recognition is incompatible with the mere modification of the dominant system of explanation, but signifies its *dissolution*. Fleck speaks here of "'mutations' in thought-style," and Kuhn, translating Fleck's concept as "revolution," sets such situations at the center of his exposition.[14] In this connection, it must be emphasized that the discoveries that shatter the valid framework do not themselves yet constitute a new thought-style or a new paradigm. Even if they prove indisputable, they remain at first "singular phenomena" on which projects for new paradigms have yet to crystallize. Kuhn speaks of "recognized anomalies whose characteristic feature is their stubborn refusal to be assimilated to existing paradigms" and which "give rise to new theories."[15] In *De Motu Cordis* (1628), Harvey described and demonstrated the actual motion of the heart and blood; yet however plainly his exposition unhinged Galenic physiology, it was far from offering its own explanation of the phenomenon, that is, a new physiological system—much less a mechanistic one. If, for example, F. Garrison regards the *De Motu Cordis* as the "starting point of purely mechanical explanations of vital phenomena," he is setting this further development too early by one significant stage.[16]

The Australian historian of medicine B. Mowry also fails to recognize the essential difference here when he rejects Kuhn's conception of scientific change by appealing to the example of the discovery of the circulation. Harvey's model, he contends, did not prevail in an "irrational," discontinuous process, but through a demonstrative procedure that was compelling according to scientific criteria of rationality.[17] This may be entirely correct for the circulation, but not for the different interpretations that it permitted. The circulation was recognized by physiologists and physicians as a singular phenomenon only when the prospect arose of a new paradigm that would make this phenomenon *intelligible* and make the physiology of the organism again a significant whole. Not without reason do we find among the first sponsors of the new model, Descartes and Robert Fludd (1574–1637), each of whom could integrate it into the new perspectives they had already developed: mechanistic in the one case and mystic-alchemical in the other. Harvey himself, who suggested a vitalistic explanatory framework as early as the *De Motu Cordis*, must also be included here. These are in fact the three essential thought-styles that were competing in the first half of the seventeenth century to be the successors of Galenism; but this "contest of . . . fields of view" ("Streit der Gesichtsfelder") was not decided by the criteria of demonstration that held for the model of the circulation itself.[18]

Moreover, only the distinction between the discovery and its reception or interpretation in its historical context explains the phenomenon that often, on closer analysis, the apparent founder of a thought-style cannot be assimilated to it and, indeed, appears to be, subjectively, thoroughly averse to it. This holds for Harvey and the further mechanistic-atomistic development of the physiology of the circulatory system. It also holds later, in the opposite direction, for Albrecht von Haller (1708–1777) whose discoveries were celebrated by his students and followers throughout Europe as the reestablishment of vitalism while he himself remained basically a mechanist and confronted this development with anything but understanding.[19]

Hence, Ludwig Fleck also compares the scientific discoverer with Columbus, who was seeking India, but found America.[20] And one might carry Fleck's analogy further: not Columbus, but Amerigo Vespucci, one of its later discoverers, gave to the new part of the globe, only then recognized as such, the name that incorporated it into a new world picture. This incorporation is beyond the control of the discoverer; even should he contribute to it, it is only the community that makes the real decision.

As to the further development after a "revolution," Kuhn advocates a strictly monoparadigmatic interpretation: "The decision to reject one paradigm is always simultaneously the decision to accept another."[21] An *extended* period without a paradigm or with a competition of claimants to paradigm status is inconceivable. This consequence follows from Kuhn's basic definitions: The "scientific community" and "normal science" are constituted precisely by a unitary, generally binding paradigm. The coexistence of different points of view, on the other hand, corresponds to what Kuhn calls the "prescientific" phase of a discipline and is overcome with its entry into the genuine process of science.[22]

Here, too, Fleck's conception proves to be broader. For one thing, he recognizes that "thought-style" is a somewhat idealized conception, obtained by a process of abstraction: for the thought-collective consists of individuals, and "The individual life of the human spirit contains incongruent elements. . . . One individual belongs to several thought-collectives, . . . which sully the purity of every doctrine and every system."[23] Further, there are the "pre-ideas," which, through their multivalence, function as connecting links between opposing thought-styles. Carrying Fleck's line of thought further, one could see in them something like "Trojan horses," which always carry along with their obvious connotations some contents foreign to any particular thought-style. These can come to light unexpectedly in a sudden switch of meaning (such as the vitalistic import of Haller's "irritability") or, on the other hand, they can be employed, through purposeful reinterpretation, in the construction of a new thought-style out of the concepts of the old one as we shall see in the case of Descartes's use of elements of Galenic physiology.

Thus, in Fleck's account, it becomes clear that both the thought-style and paradigm concepts abstract to a certain extent from the latent richness of different perspectives and from any interconnections between them. Neither Fleck nor Kuhn, however, has concretely envisaged the possibility of the *overlapping* or *interweaving* of opposing thought-styles over a longer period of time. But just such a case seems to us to occur in the development of circulatory physiology after Harvey's discovery. We shall try to show *that the machine paradigm developed by Descartes was indeed superimposed on Harvey's vitalistic conception of living things so thoroughly as to make it unrecognizable, but that the Cartesian view was not able to replace completely that of Harvey so that the latter also remained effective, as it were, in the form of a subterranean countercurrent.* That countercurrent is manifested, not only in the isolated occurrence of vitalistically oriented physiologists, but also, and particularly, in vitalistic ingredients to be found in mechanistic physiologies, especially in concepts that are "out of character," in other words, that do not fit into the new conceptual system in virtue of their traditional nature. Thus, in its individual manifestations, the prevailing line of thought proves to be "contaminated" by incongruent elements. Let us characterize this overlapping relation between two points of view by the terms "dominance" and "latency"; by this, we mean to indicate that under favorable conditions at some later time (a thought-style crisis, the exhaustion of a research program, a general change of worldview, or the like) the latent viewpoint may once more be activated and even gain ascendency.

It is our intention to consider from this perspective the relation between vitalistic and mechanistic thought from the period following Harvey until the middle of the nineteenth century. (The magical-hermetic interpretation of the circulation represented by a writer like Robert Fludd can here be given only peripheral attention; the demonstration of its latent persistence would demand a separate investigation.) We shall see how, in a relatively short time around 1750, this relation is reversed. In its turn, the mechanistic point of view now enters a latent stage, and only in the nineteenth century does it return to its ascendancy. This periodicity does not in any way signify merely a reciprocal succession or repetition. On the contrary, whatever thought-style is dominant at a given time shows itself in many ways to be influenced and modified by its predecessor. Thus it is possible to speak of an *interaction* of points of view in the course of history—an interaction that is manifested in the particular elaboration of these points of view.

From a Kuhnian perspective, the development from about 1850 appears as the definitive transition from the prescientific to the "mature" phase as the "take-off" of physiology, so to speak, or of scientific medicine as a whole under the banner of a unitary and binding paradigm.[24] But according to our conception, on the contrary, monoparadigmatic science—for Kuhn science as such—represents a *limiting case*, an ideal to which the modern natural sciences do indeed aspire, but which they nevertheless can never wholly attain. In other words, even in physiology from 1850 to the present it will be possible to demonstrate the latency of vitalistic ideas. Granted, this demonstration cannot at once claim universalizability, but the following reflection, in fact, leads to the same outcome: even if viewpoints contradicting the reigning paradigm no longer ever occurred explicitly (but every science has its outsiders!), nevertheless, the reigning thought-style would always carry in its concepts and pre-ideas the germ of another way of thinking. Otherwise, there would, in fact, be no way of explaining how it could ever come to revolutions and changes of paradigm in the history of science—unless they were interpreted in the sense of a *radical* discontinuity as each time a "creatio ex nihilo" of the new. In this way, however, the identity of the sciences would be dissolved, an identity that rests on the ultimate unity and identity of a given *object* (e.g., the human organism), and it would have to be replaced by the (questionable) social identity of the particular collectives of scientists.

Finally, as to the causes of the above historical dynamic, as we have already indicated, factors internal as well as external to the sciences have to be considered: on the one hand, above all, the consistency and plausibility of a system of thought, its "fruitfulness" in opening up a research horizon, its prognostic and practical applicability; on the other hand, the change of political-social structures, which may favor to a lesser or greater degree certain scientific conceptualizations that correspond to them, or general changes in the worldview of a culture. However, without challenging the significance of these causes for the development of the sciences, we wish, in our work, to call attention to a point of view that has received little attention, but that might gain credibility precisely from the striking periodicity of the vitalistic and mechanistic perspectives. The thesis we are suggesting is: that corresponding to this "swing of the pendulum" there is also an *ambivalence in the very object* (in this case the system of the heart and circulation) whose structure permits equally well either perspective. Further, this ambivalence would also act as a "mainspring" since the dominance of *one* point of view, by virtue of its one-sidedness (to which, according to Kuhn, scientific development inevitably inclines), would also necessarily call forth the opposing position. Thus their very monoparadigmatic tendency would itself contribute to the production of crises in the sciences. After the suppression of other points of view, the reigning view will show itself, sooner or later, to be incapable of doing justice to reality in all its aspects.

The concept of ambivalence is by no means intended to exclude the possibility that there might be more than two points of view. Still, precisely in the realm of the organism, the vitalistic and the mechanistic thought-styles appear to be related to one another in a special, in fact *complementary*, fashion. In this way, one could explain their interweaving and their reciprocal influence for all the obvious antagonism between them.[25]

Thus, according to the approach we have suggested, corresponding to a "point of view," there is also a specific "view" of its object. Instead of the concept of "thought-style" or "paradigm," therefore, we prefer the concept of an "aspect," which expresses this relation between the form and the object of knowledge. Accordingly, development of new aspects means discovering and getting to know a new "side" of the object, and it is in this sense that Descartes, too, can be called a "discoverer."

It is characteristic of Western and, particularly, of modern scientific development, that it has tolerated the contradiction between different aspects (thought-styles, paradigms) only as a passing stage to be overcome as quickly as possible. This process is still considered the normal case in contemporary philosophy of science. From its perspective, contradiction is tolerable only in a historical *sequence*; it has no value as the motor of change. From this exclusion of simultaneous plurality in the practice and theory of the sciences, it follows that each period is forced to identify the phenomena with only one or other of its aspects, namely, with that prescribed by the dominant point of view. This means, in our case, for example, the identification of the circulation with its interpretation as a mechanical system of pump and pipes. This aspect is then projected back onto the discoverer of the phenomenon, that is, Harvey. What constitutes only one aspect of the reality thus appears directly as reality itself. And yet its other suppressed aspects always continue to be present in latent form and even to be rediscovered.

In order to convince ourselves that the process presented is not the only possible one, we must look beyond the history of Western science, for example, at the development of Chinese medicine. As P. Unschuld has shown, Chinese medicine is characterized not by a linear sequence, but rather by a "growing synchronic plurality" of different systems of ideas in therapeutics.[26] On Unschuld's account, traditional Chinese thought proceeds according to a "polylinear logic. . . , which permits the *definitive* acceptance of several mutually exclusive explanatory models."[27] Unschuld also writes:

> Progress, it appears, is a European concept; the term carries the sense that one is progressing to something, but also from something, and the latter something is left behind, even though it had done good service for a while. Progress of this sort is not to be recognized in the cognitive dynamic of

> pre-modern China. In China, too, new knowledge was constantly produced, yet . . . at no time can we speak of obsolete knowledge that was definitively driven out by new items of knowledge. Not the continuous dissolution, but the retention of existing knowledge is recognizably the basic tendency.[28]

The concept of linear, cumulative progress was basic for the traditional approach to Western history of science; according to the new conception, say, of Kuhn, this progress is dissolved into a disparate sequence of self-contained systems of thought. With a glance at the history of Chinese medicine, we can now contrast, as a third possibility, the picture of a *complementary evolution*, in which the historical development of different and contradictory points of view permits the unity of reality to appear so much the more richly under different aspects. If we could reconcile ourselves to this point of view, we would have a new and significant task for the history of science: the rediscovery and reevaluation of possible points of view that have been passed over in the development of Western science, but which nevertheless survive in our culture in latent form.

II. Summary of the Argument

The subject of our investigation, as sketched above, includes the interpretations of the circulatory process in the organism developed by Harvey and by Descartes, each imbued with its own aspect as well as the interaction between the two aspects in the succeeding period. Our presentation begins with an historical glance backward at Galenic-scholastic physiology (Part B) against the background of which the difference in character between the Harveyan and Cartesian perspectives stands out more clearly. In fact, the departure of the Galenic tradition from the original vitalism of Aristotle—especially the progressive tendency since antiquity towards an externalization of the relation of body and soul and its instrumental conception—already points in the direction of Cartesian dualism; whereas Harvey's Aristotelian orientation was bound to bring him into conflict with the dominant physiological thought of his day.

In the next, longer section (Part C), Harvey's vitalistic point of view is reconstructed with special attention to parallels with and differences from Aristotelian biology. The first chapter begins by investigating Harvey's basic methodology and his conception of science. It has, as its chief goal, to overcome a dichotomy in Harvey that was suggested by his contemporaries and also prevalent in the scholarly literature: the divorce between the discoverer who works "purely empirically" and his "relapse" into medieval speculation. It will become evident that, according to Harvey's conception, the concrete living phenomenon is indeed in the first instance to be investigated and grasped as such. That, however, in order to understand it, science must in no way rest content with that task, but must seek the *principles* inherent in the phenomenon—principles of the order of entelechies. The path to this goal consists, Harvey declares, in tracing back the steps of development, that is, in moving from the organically developed and differentiated to its first and simple state: only *morphogenesis*, beginning with the fertilized egg, permits a true understanding of the developed circulation. The next series of chapters follows this retrograde path.

In the second chapter, it is demonstrated that even the *De Motu Cordis* itself cannot be interpreted in a mechanistic sense: although Harvey's exposition is concerned primarily with the concrete phenomena—in this case, the demonstration of the circulation—and is limited to a preliminary consideration of its etiology, it nevertheless already exhibits a vitalistic perspective. For one thing, the action of the heart even in its mechanical aspect appears on the scene as the imposition and extension of a primary, vital self-movement of the blood, which comes into view in its pure form in the early stages of onto- and phylogenesis. Further, in the vital heat of the blood, which is cyclically regenerated by the motion of the heart, the kinetic and qualitatively vital aspects of the circulation are inseparably bound together. In general, what is mechanical in the circulation proves to be an instrument that arises secondarily and is subordinated to the goals of the living.

As will be elaborated in the third chapter, the work on the motion of living things (*De Motu Locali Animalium*, 1628), which has hitherto received very little attention, contains essential elements of an original Harveyan vitalism. Vital motion, as Harvey here conceives it, is no longer subordinated to central control (whether by the soul or the brain), but is the manifestation of extensive autonomy and interconnecting activity on the part of the organic tissues themselves, which are endowed with self-movement. The muscles, like the heart, are "as it were autonomous living things," whose activity can be modulated and coordinated only by the brain and nerves. This new conception is reflected in the reinterpretation of the "pre-ideas" of vital heat (*calor innatus*) and of the spirits (*spiritus*); the latter, active in traditional physiology as peculiar agents in the service of the soul, become synonymous, for Harvey, with the specific capacity for movement and reaction possessed by organic substance, such as that of musculature or blood. Later, Harvey develops these thoughts further in the conception of an elementary capacity for perception and irritability in living tissue in general, and on this basis, he explains the action of the heart as the triggering of muscular contraction through the "perception" of the inflowing blood. Thus the cardiac rhythm arises from an *interrelation* of heart and blood.

The polar structure of living movement has now been described, but not yet the ultimate principle operative in it. The understanding of this is made possible only through inquiry into embryogenesis and, ultimately, into generation—both the theme of *De Generatione Animalium*, Harvey's *magnum opus* (Chapter IV). In that work, he renews the Aristotelian search for the *arché*, for the original form (*Urform*) of the heart and circulation, from which, in the last analysis, the whole organism proceeds, and he finds it, not, like Aristotle, in the heart, but in the self-moving blood. Thus, in contrast to the preformist embryology of his time, Harvey identifies the *fluid*, as yet unformed but capable of taking any form, as the germ of the organism, the stable structures of which form themselves from the motion of the blood. This primary organic substance, the "life-stuff" blood, finally appears as ensouled through a universal principle of nature in which Harvey descries the *primum efficiens* of generation as well as of the circulatory process. Summarizing, we will characterize the vital aspect of the circulation in Harvey's work with the concepts of *cycle* and *polarity*, which are interwoven as structural principles—the first borrowed from antiquity, the second, novel principle pointing to the future rather than the past.

The next part, D, presents Descartes's projection of a mechanistic view of the organism (Chapter I) and investigates the more significant physiologists who were chiefly oriented to that view, up to about 1700 (Chapter II). Even in the basic character of Cartesian science, a conception wholly opposed to Harvey emerges: Descartes acknowledges the concrete phenomena only in so far as they can be deduced from his abstract, physicalistic concept of matter. Instead of inherent principles, what governs the living as well as the dead are absolute laws of nature, which constitute a world of purely mechanical action and reaction. A consequence of this is the machine-paradigm of the organism with the essential characteristics of automatism of organ function (instead of the Harveyan autonomy), the abolition of living self-movement and its replacement by the reflex, and lastly, the separation of sensation in the body and inner feeling from the body as such and its spatiality.

Summary of the Argument 13

The physiology sketched by Descartes actualizes these basic principles above all through the conceptual reinterpretation of "vital heat" as a physico-chemical process of reaction and of the "vital spirits" as a neuronal stream of particles. In this way, the motive force of the bodily machine on the one hand and its regulation and movement on the other become explicable on a purely physical level. But the crucial connection between motive force and regulation is produced by the circulation, now conceived as a mechanical "driving belt," but also as a system of regulatory feedback. In the end, we have, instead of the vital autonomy of the organs as in Harvey, their complete subordination to the central nervous system; in the Cartesian theory of the emotions, even the heart is linked for the first time to neural control.

After the description of the mechanistic perspective, we then trace (Chapter II) the way in which this view determined the process of establishing the circulation as a fact. Descartes's influence not only contributed essentially to the ultimate recognition of Harvey's discovery; above all, the Cartesian doctrine itself appeared as a place of refuge to escape from the disorientation and skepticism that the crisis of the older paradigm had released. There emerged a horizon of prospectively fruitful research on a mechanistic-atomistic foundation, which soon produced results: notably, the theory of the regulation of the motion of the heart by the central nervous system and the discovery of the relation between respiration and "vital heat" in a reaction of chemical combustion. The price of this was the neglect of the vitalistic-morphogenetic direction of research undertaken by Harvey.

In order to document these developments in connection with Harvey's discovery, physiological writers up to 1700 will be studied with reference to their points of view and their accounts of the function of the heart and circulation. In this context, the countries most advanced in medicine will take a central place: Holland, a citadel of Cartesianism, and England, where the concealment and displacement of the vitalistic aspect by the dominant mechanistic alternative becomes especially clear, precisely in the continual invocation of Harvey combined with rejection of his own leading ideas. On the other hand, in a series of writers, the purity of the mechanistic thought-style will prove to be clouded; the heart in particular, with its mysterious rhythm and its connections with the psychological sphere seems to have resisted reduction to one aspect so that the *latency* of vitalistic conceptions comes to be expressed in "mixed" theories of cardiac action.

Finally, Part E follows the interaction of vitalism and mechanism in the period between 1700 and 1850. After a phase of increasingly challenged domination of the mechanistic aspect, Albrecht von

Haller's novel experiments on living preparations occasioned a turning point in about 1750. He epitomized his results in the concepts of an "irritability" and "sensitivity" of the organic and thus, without being aware of it, took up Harvey's conception again in modified form. The inability of the mechanistic approach to provide a satisfactory explanation of the action of the heart contributed materially to this development. Haller demonstrated conclusively the heart's independence of the central nervous system and, like Harvey, interpreted it as an autonomous stimulus-response relation between blood and cardiac muscle. Against Haller's intentions, his results were generally taken as a demonstration of the existence of specifically vital powers, which put in question the dominant dualism of body and soul. In a short time, vitalism became the new dominant point of view. It harked back to Harvey, not only in the explanation of cardiac activity, but also in the preference for a morphological standpoint. Organ functions were derived from their development and the motion of the heart, in particular, from the original self-movement of the blood in the egg; now once more, the blood was often considered the fundamental material of the organism—sensitive and itself alive.

In a further radicalization of vitalistic approaches, the physiology of the Romantic period (from about 1800) contrasted the *periphery* as the authentic sphere of living processes with the heart as mechanically acting center. The motion of the blood, too, was attributed to a "peripheral heart" to which the central heart was subordinated as a merely secondary motive force. Under the concept of *polarity*, the guiding theme of *Naturphilosophie*, which had already been introduced by Harvey but which at this point became explicit, German physiologists in particular developed various models opposing the previous conception of the circulation.

In the same way, some "romantics" now criticized Haller's explanation of cardiac motion as a stimulus-response relation, which—under the influence of the interaction of aspects!—had itself assumed clearly mechanistic features and differed little from the Cartesian reflex principle. To this conception, the idea of an autonomous movement of the cardiac muscle, independent even of the influx of blood, was now opposed. This idea was then united with the older mechanistic theory of regulation by the central nervous system—a theory which had until then remained *latent*—to produce the explanation of cardiac action through an autonomous neural center of stimulation in the heart itself (a result that was verified experimentally about 1850). Here, then, "pre-ideas" from different thought-styles coalesced to form a new conception.

Despite considerable modifications, the thought-style that

achieved a breakthrough around the middle of the nineteenth century is connected in its foundations to the Cartesian-mechanistic tradition. A survey that leads on to the present day (Part F) is intended to indicate the persistent latency of the vitalistic aspect of the circulation, in the occurrence of counter-positions as well as in elements of the dominant thought-style, which stem originally from the vitalistic rather than the mechanistic tradition. Finally, this reciprocal relation between opposing points of view is brought into connection with the *ambivalence* of the object, especially with the complex status of the heart in the organism. The historical sequence of changing perspectives thus finds its foundation not least in the very object itself, which repeatedly transcends a one-dimensional view and demands a different explanatory model.

NOTES TO PART A

1. "Mens, opinio, mores, qui nunc seculum tenent; quidni multis abhinc annis (exoletis interea omnibus, quae hodie in usu sunt) denuo recurrant?" W. Harvey, *De Generatione Animalium* (1651), *Opera* Pt. II, p. 400; trans. R. Willis, *The Works of William Harvey*, London, 1965, pp. 143–586, p. 582; cited below as DG.

2. We may take just the following quotations as typical of Harvey's fame, which is uniform in the history of medicine: ". . . (Harvey's) work has exerted a profounder influence upon modern medicine than that of any other man save Vesalius" (F. Garrison, *An Introduction to the History of Medicine*, Philadelphia, 1929, p. 246). "Harvey laid the foundations of the science of experimental physiology and experimental medicine" (C. Singer, E. Underwood, *A Short History of Medicine*, Oxford, 1962, p. 120). "The knowledge of the circulation of the blood has been the basis of the whole of modern physiology, and with it of the whole of modern rational medicine" (Ibid., p. 121). "Modern medical progress is the direct outcome of the methods so successfully advocated and practised by William Harvey" (H. B. Bayon, "The Lifework of William Harvey and Modern Medical Progress, *Proc.Roy.Soc.Med.* 44 (1951): 213–218, p. 213).

3. So, it is for example in Singer/Underwood (". . . his work soon passed into oblivion," p. 138) or in Garrison who spares only a few words for Descartes.

4. C. Bernard, *Leçons de pathologie expérimentale*, Paris, 1872, p. 481.

5. W. Pagel, *William Harvey's Biological Ideas*, Basel, 1967, pp. 28 ff., 103 f; Pagel, *New Light on William Harvey*, Basel, 1976, pp. 13–33; Pagel, "The Reaction to Aristotle in 17th century biological thought," in E. Underwood, ed., *Science, Medicine and History*, Oxford, 1953, pp. 489–509.

6. "The view that it was opposition to Aristotle which ushered in

and was largely responsible for the rise of modern science in the late 16th and in the 17th centuries is one-sided and misleading. At all events it doesn't apply to biology and medicine." (Pagel, 1967, p. 235). P.R. Sloan also makes similar statements: "If traditional Aristotelian Realism was rejected as the route to truth about nature by the physical scientists, this was by no means the case with the developing new science of the anatomists, botanists, embryologists and natural historians of the period." (P.R. Sloan, "Descartes, the Sceptics, and the Rejection of Vitalism in 17th Century Physiology," *Stud.Hist.Phil.Sci.* 8 (1977): 1-28, p. 7). But exact sciences like astronomy or physics also had not so very "exact" founders: just think of Kepler's music of the spheres or of Newton's cabbalistic-alchemical inclinations.

7. Cf. e.g. C.H. Schultz, *Das System der Circulation*, Stuttgart, 1836, p. 3; L. Manteuffel-Szoege, *Über die Bewegung des Blutes*, Stuttgart, 1977, p. 53. On the other hand, in the past century, J. Oesterreicher had already called attention to the significance of the Cartesian influence on the reception of the circulation: ". . . (Harvey's) doctrine gained recognition and rejoiced in well-deserved applause. It may well be, that this is to be ascribed, not only to the convincing facts, from which the doctrine sprang, but even more to the circumstance that it gave a rich field and broad scope . . . to those researchers who, following *Descartes*, valued atomism." (*Versuch einer Darstellung der Lehre vom Kreislauf des Blutes*, Nürnberg, 1826, p. 2).

Our chief thesis is also supported by A.C. Crombie: ". . . it is a tribute to the power of Descartes's theoretical genius that the question of vitalism and mechanism continued until the 20th century to be argued (sometimes unconsciously) in the philosophical terms established by him and his 17th-century critics." (A.C. Crombie, *Medieval and Early Modern Science* [2d revised version of *From Augustine to Galileo*, 1952], Garden City, N.Y., 1954, vol. II, p. 244).

8. See T.S. Kuhn, *The Structure of Scientific Revolutions*, Chicago, 1970, pp. vi–vii. On what follows, see L. Fleck, *Entstehung und Entwicklung einer wissenschaftlichen Tatsache (1935)*, Frankfurt, 1980; Fleck, *Genesis and Development of a Scientific Fact*, Chicago, 1979. See also, Fleck, *Erfahrung und Tatsache*, Frankfurt, 1983.

9. Fleck (1980), pp. 70, 85; Fleck (1979), pp. 51, 64. In what follows, references to *Genesis and Development* are given first to the German (G) and second to the English (E).

10. Ibid., G. p. 35 f, 172; E. pp. 25, 130.

11. Ibid., G. p. 121 f; E. p. 92 f.

12. Ibid., G. p. 105; E. p. 78.

13. Fleck (1983), p. 93; cf. also Fleck (1980), G. p. 122, E. p. 92: "All empirical discovery can therefore be understood as the supplement, development, or transformation of the thought-style."

14. Fleck (1980), G. pp. 38, 124; E. p. 28; cf. p. 94.

15. Kuhn (1970), p. 97.

16. Garrison, p. 247.

17. B. Mowry, "From Galen's Theory to William Harvey's Theory:

A Case Study in the Rationality of Scientific Theory Change," *Stud.Hist. Phil. Sci.* 16 (1985): 49–82.

18. Fleck (1980), G. p.67 n. 40; E. p. 174.
19. Cf. R. Toellner, "Mechanismus—Vitalismus: Ein Paradigmenwechsel? Testfall Haller," in A. Diemer, ed., *Die Struktur wissenschaftlicher Revolutionen und die Geschichte der Wissenschaften. Symposion der Gesellschaft für Wissenschaftsgeschichte 8* (10/5/1975 in Münster), Meisenheim, 1977, pp. 61–72.
20. Fleck (1980), G. p. 91; E. p. 69.
21. Kuhn (1970), p. 77.
22. Ibid., pp. 20 ff.
23. Fleck (1980), G. pp. 60 ff.; E. pp. 44 f.
24. Cf. T.S. Kuhn, "The Essential Tension: Tradition and Innovation in Scientific Research," in C.W. Taylor, ed., *The Third University of Utah Research Conference on the Identification of Scientific Talent*, Salt Lake City, 1959, pp. 162–74; cf. esp. pp. 168 f.
25. On this, cf. R. Löw, *Philosophie des Lebendigen*, Frankfurt, 1980, pp. 13 f.
26. P.U. Unschuld, *Medizin in China. Eine Ideengeschichte*, Munich, 1980, p. 12.
27. Unschuld, "Gedanken zur kognitiven Ästhetik Europas und Ostasiens," in *Jahrbuch der Akademie der Wissenschaften zu Berlin*, Berlin, 1989, pp. 352–66, p. 358.
28. Ibid., p. 364.

B

THE GALENIC PARADIGM AND ITS CRISIS

Summary. Galenism forms the historical background for the confrontation between vitalism and mechanism that begins in the seventeenth century. In the course of development since antiquity, the original basis of Galenism in Aristotelian natural philosophy was often modified and transformed; thus there arose an extraordinarily complex system of differing explanatory principles and interpretative levels for the physiology of the organism. In 1544, this system, which had remained essentially unchanged until the modern period, was once more presented in comprehensive fashion in Jean Fernel's physiology. Part B offers a survey of the material-somatic-psychic hierarchy of Fernel's system. Of special significance in this context are, on the one hand, the doctrine of the "spirits" as an important pre-idea for the later modern physiologies and, on the other, the traditional view of the function of the heart and circulation.

In addition to this account, various factors are pointed out that contributed to the crisis and downfall of Galenism. Reductionism in particular, to which the modern sciences are inclined, played an essential role. In the place formerly occupied by the hierarchy of explanatory principles in Galenism, Harvey, for example, sets the vitalistic aspect of the organism, harking back to its Aristotelian foundation. In contrast, Descartes extracts the material-mechanistic level from the Galenic hierarchy and declares it, from now on, the only real explanatory principle in the realm of physiology.

History, it is acknowledged, gladly permits the apparent high points of its achievements to carry within themselves the germ of crisis or downfall. Political and cultural phenomena as diverse as, say, the Alexandrian empire or classical physics around 1900 were equally affected by this principle. Such a fate also overtook Galenism, which around 1600 still represented a physiological-medical system of impressive completeness and universality and yet fifty years later had already lost its authority. A point of view that today seems to explain everything is most certain to be relativized or superseded by tomorrow.

With respect to Galenism, even more than in the fields, for example, of philosophy or physics, where Aristotle represented the foremost authority for scholastic learning, we must take care not to identify this system with Aristotelianism or to draw a straight line from Aristotle to Galen and on through the Arab physicians to Jean Fernel in the sixteenth century. Even in Galen's own system, Aristotelian, vitalistic natural philosophy represents only *one* pillar, and if the Aristotelian concepts and principles were passed on from generation to generation, still this happened only with essential displacement of their meaning and their makeup. The recurrent attempts to systematize the doctrine did not take place without influences from new, non-Aristotelian bodies of thought, which formed Galenism into a wholly unique structure in the canon of traditional learning. It is only in the light of this differentiation that Harvey's explicit acknowledgment of Aristotle can be appreciated, an acknowledgment that was in fact rather unusual for a physician of that time and somewhat contentious in character.[2] A medical faculty formed in the Aristotelian mold, as at Padua, where Harvey had visited, was rather an exception than the rule.[3]

Galen himself, in his syncretic doctrine, had already modified the Aristotelian view of living things considerably under the influence of the ancient medical tradition. This was clear above all in the conception of the soul. Aristotle had viewed it as "the full actualization of a potentially living body," and thus as that which constitutes its form and its vitality; the manifestations of the soul, the vegetative, sensitive, and intellectual, were "different only in concept."[4] In Galen, body and soul move more strikingly away from, or even against, one another; the soul is reified: its three basic functions become "parts of the soul."

The category of *finality* also undergoes a change. In Aristotle, it meant, above all, the intelligible ordering of the organism in which the parts or "organs" are related to the whole in the fulfillment of their respective functions and thus themselves contribute to the whole in their comparative autonomy.[5] In Galen, to the contrary, Stoic influences become apparent in a "universalization of the notion of end."[6] Every part of the body is now—sometimes simplistically—given its entire function in relation to the good of the soul and has its position, form, and activity thus explained. In this way, there arises an exhaustively contrived structure of means and ends, which destroys the autonomy of the organs. In the course of this development, the most diverse psychic "powers" (*dynámeis/facultates*) insert themselves between the soul as final cause and the material causal agencies, powers that are coordinated with every physiological process, duplicating them, as it were, on the psychological level. The many-leveled Aristotelian concept of *dýnamis*, primarily the "potentiality of a movement or alteration," is materialized— the *dynámeis* become *forces* by means of which the soul "operates upon the body."

These tendencies were further strengthened with the extremely exact codification of Galen in Arabic medicine.[7] In particular, the axiomatic-deductive "Canon" of medicine of Avicenna (980–1055) and the *System of the Healing Art* of Averroës (1126–1198) exercised a considerable influence on scholastic Galenism, primarily in the sense of a further reification and materialization of the relations between body and soul. Finally, Christian theology gave the element of independence and superiority of the soul even in the medical sciences a significance which could hardly be united any longer with an Aristotelian, chiefly vitalistic conception of man.

In Jean Fernel (1497-1558), all these influences converged once more in a medical system of a degree of universality never before achieved—his "Universa Medicina" appeared in 1544—a system, which in its complex sequence of stages from the rational, immortal soul down to the organs, humors and elements mirrored, as it were, the hierarchy of the medieval cosmos, of the divine and angelic powers. The multiplicity of increasingly materialized links in this chain should not prevent us from recognizing that through them body and soul are not so much mediated as set over against one another in an external-dualistic manner. Fernel's system could also be called "psycho-materialistic." Since, to a certain extent, it represented the "Summa" of Galenism and thus a presupposition for Harvey and Descartes, we need to look more closely at some of its basic traits.

I. Fernel's Summa of Galenism

In his "Universa Medicina," Fernel starts out from the elementary components of the body and then assembles the organism by introducing ever higher principles and psychological functions.[8] This order of presentation, which does not begin with the primary psycho-physical unity of the human being, but introduces and appends teleological and psychological categories axiomatically, already points, from its very foundation, to the further development in the direction of Cartesian dualism. In this recapitulation, however, we shall follow the opposite direction and then discuss some particular physiological questions.

At the summit of the hierarchy stand the three functions of the soul: the *facultates animae rationalis* or *animalis, vitalis,* and *naturalis* (i.e. the rational = intellectual, the vital and the nutritive = vegetative soul). With these are coordinated the organ systems of brain and nerves, heart and arteries, liver and veins, respectively. (In Galen, the triad plays a major role as organizing principle). As the highest goal, the soul is at the same time the real cause of bodily processes, which are not explicable in themselves, that is, in terms of movements and forces on the material level.[9] For this purpose, there is always a need for a great variety of "sub-faculties," which are subordinated to the three highest as "instruments"; "for so many achievements, so many capacities."[10] Thus the *facultas naturalis* has four auxiliary capacities: the attractive, expulsive, restraining, and concoctive (*facultas attrahens, expellens, continens, concoquens*). Many capacities are specifically tied to individual body parts as the *vis motrix* or *pulsifica* are to the heart and arteries. However, the faculties as such, that is, as psychological categories, cannot act directly on the organs and body parts; rather, they are represented in them by a special, very subtle substance, the *spiritus*, or spirits.

a. COROLLARY: THE DOCTRINE OF THE SPIRITS

In the spirits, we meet an idea that had maintained a firm and significant place in physiology since antiquity. In concepts like that of "nerve fluid" or of the vitalistic "life-stuff," it continued its influence into the nineteenth century; it was only gradually that extra-biological, electrochemical explanations eventually took its place. Theories like those of phlogiston or ether, which in the last analysis had a common origin, developed on the same lines, that is, under the assumption that such intangible or invisible phenomena as "life," "heat," or "light," which clearly penetrated the most diverse bodies, must be bound in their diffusion to subtle, ubiquitous media or imponderables.

From the beginning, *air* and *fire* were the two elements in which the medium of life had been sought. This polarity is visible for the first time in the Ionian natural philosophers Anaximenes and Heraclitus. In the ancient medical schools (above all the Sicilian, later the pneumatic), there emerged from the two traditions, on the one hand, the concept of the *pneúma* (the later *spiritus*), which distributed itself in the body through the blood, as the breath-like carrier of life, sensation, even consciousness, coming from the air we breathe; on the other, the notion of "innate heat" (*émphyton thermón*, later *calor innatus*) as the source of the vitality and self-motion of living things, mirroring within themselves, as it were, the sun as principle of all cosmic movement.[11]

Both lines of thought were overlaid on one another in many ways, and so in later developments *spiritus* and *calor* are not always clearly differentiated. Yet heat remained predominately associated with the vegetative-vital aspect of the living being, and the spirits, on the other hand, rather with the mobile and sensitive, "animal" organism—and, not without reason, the latter persevered longest precisely in the physiology of the nervous system. Granted, the *spirits* had been related to the nervous system only since the third century B.C., when the Alexandrian anatomists distinguished the nerves from the sinews. Hence it was the *blood* as ensouled life-stuff, present everywhere in the body that was, in the first instance, the sole carrier of the spirits as well as of the innate heat. In this connection, therefore, the *heart*, situated in the direction of the endpoint of respiration and at the center of the motion of the blood, acquired increasing significance for the concepts of physiology.

Let us simply emphasize the conceptions that were given concreteness in the course of history by mentioning some particular points.

Diogenes of Apollonia, philosopher and physician of the fifth century B.C., assumed that the innate heat was indeed already present in the embryo, and thus resided in "vegetative" life, but that only post-natal breath lent to it the higher forms of soul (sensation and consciousness). Moreover, in the Hippocratic treatise *On the Sacred Disease*, the pneuma appears as a derivative of inspired air, which first evokes consciousness in the brain before it flows with the blood into the rest of the body and animates it.[12] While the concepts of the primacy of air as the stuff that animates belong to the tradition of Anaximenes, for Empedocles (third century B.C.), under the influence of Heraclitus, the innate heat becomes the first source of life, sense, and consciousness. It is tied to the blood, but concentrated in the heart, which also arises as the first organ in embryogenesis.[13]

The influence of this doctrine reaches far. According to the later Hippocratic treatise *On the Heart*, but also in the view of Plato and of Aristotle, the pneuma is indeed again the carrier of psychological functions; however, it no longer derives from the air we breathe, but from the blood that swells and vaporizes through the *heat of the heart*—in keeping with the everyday example of milk that boils over when heated. The heart now becomes more and more the origin of "surges" of feeling and—in Aristotle—the central organ of sensation and perception, which is mediated through the pneuma. In contrast, respiration acquires a mere cooling and regulative function vis-à-vis the heart, which would otherwise overheat.[14]

Finally, Galen forms a clever synthesis of the two lines of thought. He leaves to respiration the tempering function, but now gives it access to the heart through the pulmonary veins. There it meets the blood formed in the liver, and now the fire in the heart produces out of *both* of these the *zotikon pneuma*, the later *spiritus vitales*, carriers of heat and vitality. From this there arise is the brain the yet subtler *pneuma psychikon* (=animal spirits), which transmits the higher functions of the soul through the nerves.[15] Here, research into the central nervous system in the Alexandrian school (Herophilus, Erasistratos, third century B.C.) comes to a close; the heart must resign a part of its higher functions to the brain, but remains the center of emotional life.

As is evident, Galen found in all points a compromise between the "Anaximenian" and "Heraclitean" positions; blood, inspired air, and cardiac fire all contribute equally to the formation of the medium of life. To be sure, the price of this consists in a complicated anatomical construction: air is brought to the heart through the pulmonary veins; on the same path, but in the opposite direction, sootlike substances that necessarily arise through the combustion of blood in the heart are supposed to depart. For this purpose, the mitral valves, unlike the other cardiac valves, must be thought of as partially porous. Finally, as is well known, Galen postulates the tiny openings in the septum of the heart through which finer blood that has been filtered can enter the left chamber as appropriate fuel for the fire of the heart. Thus the pulmonary veins can no longer be considered the place where inspired air is metabolized. What brought Galen's system into difficulties in the modern era—and finally led to its collapse— were the assumptions that led to such necessary anatomical inferences and not its basic physiological conception, which had for a long time represented an acceptable combination of originally contrary views. With the abandonment of this system, as we shall see later, the old polarity between "Anaximenian" and "Heraclitean" conceptions also reemerged.

The distinction between different *spirits* made by Galen did in fact follow from the differentiation of the functions of the soul in Plato and Aristotle, yet it meant, at the same time, a renunciation of the "life-medium," which had at first been considered unitary. Now, in its place, diversely materialized and localized *spirits* stepped in as "instruments" of particular functions of the soul. In this way, the tendencies of post-Aristotelian physiology that we have already mentioned became more marked: the reification of the soul-body relations and their detachment from their original bases in experience, which were replaced more and more by theoretical constructs.

In Fernel, to whom we now return, we find a doctrine of the *spirits* worked out in detail. From nutriment, the liver forms *natural spirits*, which spread through the venous system and mediate the functions of the vegetative, nutritive soul.[16] From blood and inspired air there arise in the left heart the finer *vital spirits*, which distribute central heat and vitality in the body through the arterial system. From these, the yet finer *animal spirits* are formed in the brain; they fill the cranial ventricle and stream from there through the nerves (which, since Galen, had been generally thought of as pipes) into the sense-organs and muscles in order to effect sensation and movement. Equally at home in the fluid and solid mediums, the *spirits* fill the space of the body; as *vinculum animae,* they transfer the activities of the *faculties* to its elementary components so that the latter are subordinated to the teleologically organized whole.[17]

Those elementary components of the body are not of atomistic, but of qualitative nature, that is, dry, warm, cold, or moist; pairs of these qualities form the elements of fire, water, earth, and air. Here, indeed, we have arrived at the level of the genuinely effective causes. Galen, too, had enunciated the view that "bodies affect one another actively and passively in virtue of the warm, cold, dry, and moist."[18]

In this context, innate heat is of central importance for physiological processes. Under its influence, the assimilation of alien matter is completed in a series of stages. One instance of such penetration by bodily heat, for example, traditionally called "concoction," occurs in the stomach and liver, a process by which solid material—earthy nutriment—is transformed into fluid, that is, into venous blood. This blood flows from the liver, as its central organ, into the periphery, where, through further cooking, it becomes "solid" again, but now as flesh, and thus at a higher, organic level. Heat, on the other hand, is the instrument by which, in the left heart, the *vital faculty* assimilates air to the organism in the form of *vital spirits*. The blood also enters into this process; it is introduced from the liver through the pores of the septum and vaporized through the heat of the heart.

Thus, while the solid material, the "earth," is distributed through the veins, the "air" finds access to the body through the arterial system. Only at the periphery do the two meet and in so doing serve as the basis, on the one hand, for the development and nourishment of the tissues and on the other, for their vitality and mobility.[19] At the same time, the fluid (or watery element) serves as "solvent" or transitional stage with heat (or fire) serving as catalyst for the process of transformation of air and earth.

Thus this physiology was able to integrate in an impressive manner all the basic elements of the dominant doctrine of nature; admittedly, it was thus indissolubly bound to the Galenic interpretation of the functions of the heart and blood.[20]

Its chief characteristic is the juxtaposition of the two blood systems in which, at the same time, a centrifugal direction of flow predominates. This means not only the continual formation of new blood in the liver and its oozing away at the periphery; it also shapes the character of the cardiac-arterial system, which is subordinated to a unique function much more directly than in modern physiology, namely that of respiration. In *De Usu Partium*, Galen treats the heart under the theme of breathing and compares it with a bellows for its active motion is that of *diastole* through which it sucks air as well as blood into the left ventricle.[21]

Thus the heart is the real respiratory organ; the lung serves only for the initial preparation, refinement, and warming of the air. The reason for this lies in Galen's conception of "innate heat": while Aristotle tried to distinguish this from ordinary "physical" fire, Galen assumes such a fire in the left heart, which needs the blood as fuel as much as it does ventilation and cooling through the breath as well as the removal of waste material through the pulmonary veins.[22] Fernel accepts this Galenic physiology of *calor* and with it the complicated flow and valve mechanics in the region of the heart.[23]

According to Galen and Fernel, the heart does not belong to the muscles since these are all subject to voluntary motion. Instead, the rhythmic heartbeat arises through a special *faculty* of the vital soul, the *vis pulsifica*; it spreads out from the heart to the arteries. Their pulse, too, is a respiratory movement. With their expansion, the arteries suck air through the skin into the body for cooling and ventilation; their contraction allows excretions of the tissues to escape on the same path. "Living things are given breath and pulse for one and the same reason," as Fernel says; inhalation corresponds to diastole, exhalation to systole.[24]

Finally, the dynamic of the motion of the heart is also analogous to breathing. The blood is not so much pressed into the arteries from the heart as it is, rather, sucked from them in its diastole. The peripheral motion of the blood is again determined in the first instance by the *facultas attrahens* of the organs and tissues. Here there exist everywhere arterio-venous anastamosesi and subtle pores in the vascular system for the care of the parts, which is regulated according to their respective *attractio* (i.e., their need) or *expulsio* (purging).[25] All of this, in its totality, has as its consequence a relatively easy-going haemodynamics, which Galen compares with the watering of a terrace garden, but also with the ebb and flow of the sea: above all, in the vessels of the portal vein, the blood *oscillates*—like the movement of the breath.[26]

The local distribution of the blood, heart, and spirits, like the purging of the tissues, is thus subject to a complex network of regulating forces; arteries, veins, skin, and tissues together form an open metabolic system. Thus Galenic physiology preserves in addition to its "multipolarity" (manifested in the triad heart-liver-brain) a plainly "decentral" component as well.

II. The Crisis of Galenism

We have seen how, in Fernel's system, the juxtaposition of body and soul, with the diverse, interacting levels to which forces are assigned, leads to a plurality of explanatory principles, which overlap in a complicated way, so that in a given case, the decision whether a particular faculty-heat, spirits, humors, qualities, or elements-is to be held responsible for a phenomenon can often be made purely on the basis of tradition in which such different streams in the history of ideas have flowed together. The result is a lack of relation between phenomena and explanatory principles, an "overburdening of the phenomena." In fact, a more or less radical reduction of this multiplicity of explanations characterizes all the successors to Galenism: it was customary to invoke William of Ockham's principle: *entia non multiplicare praeter necessitatem*. This reductionism affected the heart in particular, which had previously united in itself a plurality of functions and forces. It effected, in mechanical fashion, the motion of blood and air; as the organ of respiration and metabolism, it produced spirits and arterial blood; it was the carrier of the *facultas pulsifica*, seat of innate heat, of the vital soul and, finally, of movements of emotion.[27] If we compare with this situation the heart in modern physiology, we see it reduced to the role of a mechanical organ of impulsion within a partial system of the organism, the circulation. After a merely apparent reevaluation of the heart in Harvey's case, the downfall of Galenism was soon followed by the progressive limitation of its role in the physiology and psychology of human beings.

In medical systems like Fernel's, Galenism had once more reached a high point in its power of integration and completeness. This achievement can be appreciated if we consider that, in what immediately followed it, none of the rivals for paradigm status that arose was able to take its place as the heir of Galenism. The battles for succession around the fallen empire continued into the nineteenth century, and Galenic doctrines were able to persist so long in the universities, not least because no new physiological system offered a similar power of integration. But in that case, what did really cause the crisis and fall of Galenism in the seventeenth century?

Doubtless, the discovery of various "anomalies" that could not be fitted seamlessly into the doctrine played an essential part.[28] Among these, for example, was Vesalius's demonstration of the impenetrability of the cardiac septum (1543), Michael Servetus's and Realdo Colombo's discovery of the pulmonary circulation (1553 and 1559 respectively), or that of the valves in the veins by Harvey's teacher, Fabricius of Aquapendente (1603), and finally, Harvey's discovery of the circulation itself. Yet such anomalies on their part could come into the field of vision at all only through the development of new points of view and new aspects after the late Middle Ages. Thus to a way of thinking that once more gave observation of the phenomena precedence over their inherited interpretation, Galenism was bound to appear increasingly suspect—whether such an approach appealed to Bacon or (as in Harvey's case) to Aristotle. Above all, the doctrine of animation and of form characteristic of Galenic-scholastic physiology was a persistent stumbling block for the emerging mathematical, mechanical, and atomistic perspectives on nature and man. But the investigation of magical connections between the hierarchies of the microcosm and the macrocosm in the Neo-Platonic, Paracelsian, and animistic movements of the Renaissance also stood in clear contrast to the physiology and pathology of Galenism, which was founded exclusively on the particular living being and on the individual soul.

Finally, from the other side, the dominant paradigm found itself threatened by a mode of thought that rejected complex hierarchies in general and strove instead for unification and centralization-whether in the religious, social, or scientific sphere, in the heavens, on the earth, and in man himself. We find an example of this in the person of the theologian and physician Michael Servetus. He was a convinced Unitarian and denied the trinitarian nature of God as well as the triad of life forces in the human body. The pulmonary circulation was for him above all a proof that there was only one kind of blood as carrier of one soul. But in Harvey and Descartes as well, we meet such unifying and centralizing tendencies—in parallel, we should not forget, with the emergence of absolutism in England and France. Against this trend, Galenism, with its plurality of explanatory principles, could not assert itself. In this connection, what befell Galenism was like what happened to another great doctrine of nature from antiquity, the Ptolemaic system; not the lack of explanatory power, but above all, its complexity brought it increasingly into discredit.[29]

Last but not least, however, it was precisely the inclusiveness, the completeness of the Galenic system that proved fatal to it: for it stood in opposition to the gradual development of the current worldview, even of the attitude to life, from the static to the dynamic, from insertion in the tradition to the search for change, for the novel in all areas. Discoveries presuppose, in the first place, a belief in their possibility. If the Middle Ages had accepted Galenism uncritically and wholesale as an appropriate medical appendix to the theological hierarchy, its lack of the capacity for innovation now became evident. The new sciences, for their part, cared little for syntheses between the traditional and the new conceptions or discoveries. What had been accepted as a whole was now rejected as a whole; and what seemed to be self-evident within the Galenic system now appeared as empty tautology.[30]

Thus physiology, like modern science in general, took the path we have already indicated of a monoparadigmatic development rather than of synchronic multiplicity. To the degree to which science emancipated itself from theology, it took over theology's monotheistic heritage.

Notes to Part B

1. In the view of T.S. Kuhn, Galenic-scholastic medicine would be assigned to the "pre-historic" stage of medicine, not to medicine as "normal science," which is distinguished by a "puzzle-solving" and cumulative process of research (Kuhn, 1970, pp. 23ff). On the other hand, Kuhn recognizes statics, optics, and astronomy, for example, as pre-modern paradigms. However, the concept of paradigm seems to me appropriate here on the basis of the general recognition of Galenism by the "scientific community" of (school) physicians, its canonized transmission, the existence of textbooks, and so on.

2. "... tantopere tamen apud me semper valuit *Aristotelis* autoritas; ut non temere ab illa recedendum putem" (Harvey, *DG*, p.39; W. p. 207.

3. Cf. H. Butterfield, *The Origins of Modern Science, 1300-1600*, New York, 1957, p.48.

4. Aristotle, *De Anima* II, 412 a 25ff (referred to in what follows as DA).

5. The dialectic of finality and autonomy of organs within the organism is confirmed by the analysis of P. Pellegrin, who even speaks of an "elevation of the *moria* [parts] to the level of *ousiai* [substances]" in Aristotelian biology. See P. Pellegrin, "Aristotle: A Zoology without Species" in A. Gotthelf, ed., *Aristotle on Nature and Living Things*. Pittsburgh/ Bristol, 1985, pp. 95-115, p. 107.

6. Cf. R. Löw, *Philosophie des Lebendigen*, Frankfurt, 1980, pp. 77 and 83.

7. On this, see K.E. Rothschuh, *Konzepte der Medizin in Vergangenheit und Gegenwart*, Stuttgart, 1978, pp. 199ff.

8. Johanni Fernelii, *Universa Medicina, Physiologiae Libri I-VII*, Paris, 1644; referred to here according to the sixth edition, Frankfurt, 1607. Compare further also the account of Rothschuh (1978), pp. 202ff. *idem*, "Das System von Jean Fernel (1542) und seine Wurzeln," in *idem, Physiologie im Werden*, Stuttgart, 1969, pp. 59-65; *idem*, "Technomorphes Lebensmodell contra Virtus-Modell (Descartes gegen Fernel)," *Sudh.Arch 54* (1970): 337-354.

9. "Corporis actiones nec a se nec a corpore proficisci. Actionum corporis causam esse animam" (*Universa Medicina* V. 1. p. 169). On the triad of the faculties and organ systems, see V, 14.

10. "Quot opera tot esse facultates," *ibid.*, V, 2, p. 173.

11. On this, compare F. Rüsche, *Blut, Leben und Seele. Ihr Verhältnis nach Auffassung der griechischen und hellenistischen Antike, der Bibel und der alten alexandrinischen Theologen*, Paderborn, 1930; E. Mendelsohn, *Heat and Life. The Development of the Theory of Animal Heat*, Cambridge, MA, 1964.

12. "On the Sacred Disease," esp. §§ 16 f., in *Oeuvres complètes d'Hippocrate*, ed. Littré, Paris, 1839-1861, vol. 6, pp. 350-397. Cf. also Rüsche, pp. 112 ff., 117 ff.

13. *Ibid.*, pp. 130 ff.

14. "On the heart," *Oeuvres complètes d'Hippocrates*, vol. 9, pp. 76-93; Plato, *Timaeus*, 70a7-d6, 79 d1 ff.; Aristotle, *De Resp.* 20, 480 a 2-15 (on the action of the heart and innate heat); *De Generatione Animalium* (= DGA) II 6, 743 b 26ff., III 4, 666a 10 ff. and 34 ff.; *De Juv* 3, 469 a 10 ff. (on the heart as center of perception and *sensorium commune*).

15. Cf. O. Temkin, "On Galen's Pneumatology," in *Gesnerus 8* (1951), pp. 180-189. In fact, Temkin considers not only the *spiritus naturales* of the liver, which occur later, but even the *spiritus vitales* to be not certainly Galenic.

16. *Universa Medicina*, IV, 11, pp. 166 ff. These spirits were indeed often identified with the venous blood itself.

17. "... est igitur spiritus corpus aethereum, caloris facultatumque sedes et vinculum, primumque obeundae functionis instrumentum"; *ibid.*, IV 2, p. 145.

18. C. Galenus, *De naturalibus facultatibus*, II 9, p. 126, in Galeni, Claudii, *Opera*, ed. C. Kühn, Leipzig, 1821-33, Vol. II. The corresponding sections in Fernel are to be found in Book II, 3-8.

19. We find this polarity again today in the opposition between anabolic and katabolic metabolic conditions: the processes of synthesis and composition take place above all in the *venous* hepatic system; the processes that consume energy and lead to decomposition—and that chiefly serve the higher vital functions!—take place above all through the participation of *arterial* oxygen. On Fernel, cf. *Univ.Med.*, VI 1-6 and 16-18.

20. On what follows, compare the account of E. Gilson, "Descartes, Harvey et la Scholastique," in *idem, Etudes sur le Rôle de la Pensée Médiévale dans la Formation du Système Cartésien*, Paris, 1984, pp. 51-100 (esp. pp. 55 ff.); A. R. Hall, "Studies on the History of the Cardiovascular System. 1. Galen," *Bull. Hist. Med.* 34 (1960), pp. 391-413; Pagel, 1967, pp. 127-136.

21. C. Galenus, *De Usu Partium*, VI, 2, pp. 411 ff., VI 15, pp. 481 f., *Opera*, Vol. III.

22. For example, in DGA II 3, 737 a 7 f., or in DA II 4, 416 a 15 ff: "... for the growth of fire is boundless, so long as there is something to be burned, but of all things composed by nature there is a limit and a *ratio* of size and growth." Cf. Mendelsohn, pp. 8-26.

23. Fernel, *Univ. Med.*, VI 16, 18 (esp. pp. 296 f.).

24. "Ergo unius eiusdemque usus causa respiratio et pulsus dati sunt animalibus," *ibid.*, VI 17, p. 194.

25. Fernel, *Univ. Med.*, V 4, p. 179; VI 5. pp. 246 f.

26. Galen, *De Usus Partium*, VI 10, pp. 454 f.; *De Naturalibus Facultatibus*, III 15, pp. 210 ff. (*Opera*, Vol.II)

27. According to Fernel, *Univ. Med.* V 19, pp. 214 f., the heart is the seat of emotions like fear, joy, anger, or pain; the desires, he attributes to the liver.

28. On the concept of anomaly, see Kuhn, pp. 52 ff.

29. Typical for the tendency of the time is the statement of a collaborator of Copernicus, Domenico da Novara," ... that no system so cumbersome and inaccurate as the Ptolemaic had become could possibly be true of nature"; see Kuhn, *Structure*, p. 69.

30. Molière's comedies illustrate satirically this popular notion of the time, and the ridicule of Galenism as a heaping up of empty schematisms is also gleefully referred to time and again by modern historians of medicine and biology (who ought to know better); cf. e.g. (among others) Rothschuh, 1978, pp. 207 f., or J. Needham, *A History of Embryology*, Cambridge, 1958, p. 72.

C

WILLIAM HARVEY:
THE VITAL ASPECT OF THE CIRCULATION

The history of Harvey's reception deserves an investigation on its own: seldom has a person in the history of medicine received such diverse and contradictory appraisals as has the discoverer of the circulation of the blood (even that title has not gone unchallenged!).[1]

Harvey has often been seen primarily as an empiricist and bracketed with his contemporary, Francis Bacon (1561–1626). Thus Erna Lesky advocates the view that Harvey, in common with Bacon, turned against reliance in authority and scholastic Aristotelianism; he is said to have followed Bacon's epistemological principles in his account of the circulation and thereby to have "clearly demonstrated their fruitfulness."[2] Especially in the nineteenth century, but also in the twentieth, Harvey has repeatedly been claimed for the mechanical and quantifying view of nature. "The thought appears thoroughly modern," says E. Ackerknecht, for example, in his history of medicine, "that Harvey isolated his phenomenon. He concerns himself only with the mechanical operations of the circulation. . . . Harvey's point of view was mechanistic and corresponded to the dominant attitude of his time. He analyzed man and beast as machines."[3] Similarly, we read in A. Castiglioni's history of medicine: ". . . (Harvey) understood that the return of the blood to the heart by means of the veins was a mathematically demonstrable fact. . . . It is this way of considering the circulation from a mechanical and dynamical point of view that constitutes the truly inspired part of his work. . . ."[4] The historian of biology, T. Ballauf, also comes to the conclusion: "The ancient theory was qualitative; Harvey's theory springs from quantitative considerations. . . . The essence of life in beast and man is thus comprised in a mechanism of circulation."[5]

In recent decades, on the other hand, there has been a growing tendency to do justice also to the "obscure" side of Harvey and to understand him as a critical Aristotelian as, for example, in G. Plochmann[6] or in Pagel. H. Driesch had already seen in Harvey the first post-Aristotelian vitalist.[7] Pagel accepts this estimate,[8] and C. Webster also comes to the conclusion that ". . . Harvey regarded reference to teleological and vitalistic principles as necessary for the solution of crucial problems in biology."[9] However, this side of Harvey was also, and continued to be, the object of harsh criticism—going as far as the reproach of "medieval mysticism."[10] Some works in particular, like the *De Motu Locali Animalium* or the *De Generatione*, came under the verdict of "scholastic speculation" as, for instance, in the case of J. Needham: ". . . (Harvey) didn't break with Aristotelianism as a few of his predecessors had already done, but on the contrary, lent his authority to a moribund outlook, which involved the laborious treatment of unprofitable questions."[11]

This survey should suffice to make clear the difficulty of interpreting Harvey against the background of the literature on his life and work, which is also extraordinarily extensive. There are probably two chief reasons for the contradictory views we have presented, which, particularly on certain questions (such as the significance of the quantitative method or of the analogy of the heart to a pump), acquire the character of a stubborn, occasionally vehement debate. For one thing, Harvey's work, especially his discovery of the circulation, represents a sensitive point, as it were, in the history of medicine, the interpretation of which is of far-reaching significance for our judgment of medicine up to this day. The other reason lies in Harvey himself; Pagel has expressed it best:

> . . . it would appear that there was a time when what sounds contradictory today was no contradiction. Unification of what is today sound and relevant with its apparent opposite must have been possible in the same mind, which yet somehow retained its integrity and power.[12]

In the light of this special difficulty, it seems appropriate to begin by taking another look at the methodology and the character of Harvey's science. Perhaps this will also give us some indication of how we can then approach the aspect of the circulation that he stressed.

I. The Science of the Living

Summary. *Harvey's conception and methodology of science, as he himself explained it in some passages, is oriented to the Aristotelian philosophy of science. Its special characteristic is the intertwining of inductive and deductive method: science begins with sensible givens, but, through induction, ultimately aims at the principles from which particular phenomena can be deduced; at the same time, these principles represent the real causes of natural entities. For Harvey, science is not complete without the knowledge of the first principles—a conclusion that shows the commonly proposed separation of the "philosophizing" from the "modern," experimenting Harvey to be just as inadmissible as the removal of the mechanical from the dominant vital aspect in Harvey's physiology.*

However, despite the significance of the principles, induction nevertheless remains for Harvey the indispensable foundation of science; that is, the concrete, perceptually given phenomena cannot be "falsified," so to speak, through the principles or reduced to abstract-geometrical structures—as is the case, in contrast, in Cartesian science. Rather than aiming at general laws of nature, Harveyan science seeks inherent causal principles (formal, efficient, and final cause), which guide the development of every individual living being. But these emerge most clearly at the beginning of embryogenesis. Thus the key to the understanding of the circulation, too, lies in the confirmation of its origin and development from the 'punctum saliens,' the first self-moving configuration of heart and blood.

"There is no science which does not spring from some preexisting knowledge"—so writes Harvey in 1649 at the beginning of his first polemic against Jean Riolan, the "Exercitatio anatomica de circulatione sanguinis"; and he emphasizes this once more in the second.[13] Similarly, Aristotle's theory of science in the "Posterior Analytics" begins: "All reasonable teaching and learning arises from preexisting knowledge."[14] Aristotle is the authority to whom Harvey chiefly appeals in his discussions of method, above all in the preface to *De Generatione:* "And foremost of all the ancients, I follow Aristotle."[15]

According to Aristotle, all knowledge arises from previous knowledge and understanding, but advances in the course of this process in two directions, namely, inductively or deductively: both belong equally to science (AP I, 71a5ff.). However, only the deductive process produces certain knowledge, because it originates from principles that are ". . . better known than and prior to the conclusion, and are its cause" (I,2, 71b21ff.). Aristotle examines later the way in which the first premises of a science, the *principles*, are to be found; first, he is concerned to clarify the relation between induction and deduction:

> But things are prior and better known in two senses. For it is not the same thing that is prior *by nature* and *for us*, and better known by nature and for us. I call that prior and better known for us, which lies closer to *sense perception*, and prior and better known without qualification that which lies further from perception. Farthest from it are the *most universal*, nearest, the *particulars*. And these are opposite to one another. (I,2, 71b33–72a5; author's italics; see also *Phys.* I,1, 184a16ff.)

Granted, in its *genesis,* our knowledge has sensible phenomena as its starting point and arrives ultimately at the principles; nevertheless, from a *logical* point of view, it must take the reverse path, from what is prior "by nature," or objectively, to the individual phenomenon whose *real causes* it grasps in the principles. The same intertwining is found, according to Aristotle, in the development of a natural entity itself: all becoming proceeds from a substratum in which the form is led from potentiality to actuality.[16] But while, from an onto*genetic* perspective, the actualized structure stands at the end of the development, it is nevertheless, as the goal of development (entelechy), the genuine reason for the coming to be and thus onto*logically*, in the hierarchy of being, it is prior to the particular stages of development or to the parts of which the organism is composed.[17] So human knowledge replicates the natural process of becoming, and, indeed, in both directions; it is precisely in this way that it becomes adequate knowledge of reality.

Thus, despite the logical precedence of deduction, human knowledge depends on induction:

> It is impossible to come to know universals except through induction . . . But it is impossible to be led to perform induction if we lack perception. (AP I 18, 81b2).

It is no different with the highest premises, the very principles of a science. Since, as principles, they can neither be derived nor demonstrated from something else, nor, on the other hand, recollected as innate, as Plato taught, there remains here too only a special inductive path to the knowledge of them. It rests on the power of the mind to extract the universal step by step from the impressions of the senses. In this way, first, from persisting impressions there arises memory; next, from many memories of the same kind of thing, we obtain *experience*—which is presupposed, if, finally, we are to reach the principles in a process of intuitive knowledge resulting from continual reference back to sense perception (AP II 19). Thus, although the universal cannot be recognized immediately in a single act of perception (I 32, 87b30f), it is nevertheless contained in it potentially and can be brought to light in the way just described.

> We see, then, that we must learn to know the first principles through induction. For this is also how sense perception implants the universal. (II 19, 100b3ff.)

Harvey, too, recognizes on principle the superiority of deductive science, but stresses even more its dependence on induction and sense perception. In the preface to the *De Generatione*, where Harvey explains his method of research, we read:

> Although there is but one road to science, that to wit, in which we proceed from things more known to things less known, from matters more manifest to matters more obscure; and universals are principally known to us, science springing by reasonings from universals to particulars; still the comprehension of universals by the understanding is based upon the perception of individual things by the senses.[18]

There follows a quotation from Aristotle's *Physics* and a paraphrase from the *Posterior Analytics*, which conclude with statements that, as Harvey remarks, are to be understood in complementary fashion: "whence it is necessary to proceed from universals to particulars" (*Physics*); and "whence it is advisable from singulars to pass to universals." (*Anal.Post.*)[19]

What Harvey writes here and the passages he quotes are not for him mere abstractions. His science consists *equally* of the path to the principles and the deduction from them; he recognizes and applies in his works both induction and deduction. In the *De Generatione* itself, the detailed investigation of generation and development is followed by the stepwise search for the principles that guide these processes from which, in turn, conclusions follow, even explicit syllogisms (DG 402/584f.) *De Motu Locali Animalium* proceeds, for the most part, from known and traditional principles of living movement, which are tested in the light of biological and medical experience. In the *De Motu Cordis et Sanguinis,* the structure is at first sight not so evident.[20] Here Harvey switches constantly between inductive or deductive arguments for the circulation, on the one hand, and on the other, the use of the circulation as a *premise* from which, and only from which, certain phenomena can be explained. Admittedly, he consciously abandons the search for the *highest* grounds and principles for the circulation, but nevertheless employs the form of deduction for his demonstration, as, for example, in Chapter 9, where he uses a disjunctive inference:

(1) A fluid flowing constantly through a point (namely, the heart will either (a) be continually formed anew (namely, from nutriment) or (b) move in a circle.
(2) Possibility (a) is excluded for quantitative reasons (the quantity of nutriment is insufficient for new formation);
(3) Hence, the fluid moves in a circle.

In the same chapter, the conclusion from the premises serves in turn as a presupposition from which certain knowledge about particular phenomena can first be deduced (for instance: "And now the cause is manifest, wherefore in our dissections we usually find so large a quantity of blood in the veins, so little in the arteries. . . .").[21]

Finally, the central passage in the book as well, in which Harvey reports his discovery itself, goes in both scientific directions (DMC 55f., 45f.). Numerous observations of the motion of the blood and the structure of the heart, inspections of its symmetry and function formed the foundation. To this were added, Harvey continues, repeated reflections and attempts at solution (*cum . . . saepius mecum et serio considerassem . . . animo diutius evolvissem*). Harvey describes exactly the difficult, inductive-intuitive path of the discovery of the universal from particular observations; until finally ". . . I began to think whether there might not be *a motion, as it were, in a circle*. Now this I afterwards found to be true" (DMC ch. VIII, p. 46) (*coepi egomet mecum cogitare, an motionem quandam quasi in circulo haberet; quam postea veram esse reperi*. Caps. in W.). In this way, the point of reversal is indicated: the circulation, which does not represent any highest, indemonstrable principle, must now itself be demonstrated from premises in order to be counted as certain knowledge; the mere agreement with observation is not enough. Harvey does not conceive the circulation in the modern sense as a "model" for which it is only a question of agreement with universal laws of nature on the one hand and careful protection against falsifying observations on the other; rather, he regards it as inherent in the genetic and logical order of nature, as subordinate to its principles. But it is these principles that also guide the scientist on his way.

Let us summarize what we have said so far. Harvey's understanding of science cannot be restricted to a determinate approach. It is the particular starting point in the network of natural interconnections, whether selected or presupposed, that suggests to the scientist either the "ascending" or the "descending" direction. Yet total science needs the careful, intuitive procedure in search of the principles as much as the application of those principles for the derivation and confirmation of knowledge; but most of all, it needs the constant testing of knowledge through observation and experiment—"avoiding that specious path on which the eyesight is dazzled with the brilliancy of mere reasoning, and so many are led to wrong conclusions. . . ."[22]

Harvey emphasizes the last point repeatedly, not only against the scholastic reception of Aristotle, but also against the emerging new rationalism as embodied in Descartes.

". . . all true science rests upon those principles which have their origin in the operation of the senses (*scientia omnis perfecta, iis principiis innititur, quae ex sensu compertis originem ducunt;* DG Pf; W. 158): this proposition can stand as a summary of Harvey's position—as can the passage quoted in connection with it from Aristotle's *De Generatione Animalium*:

> And when the investigation shall be complete, then will sense be rather to be trusted than reason; reason, however, will also deserve credit, if the things demonstrated accord with the things that are perceived by sense.[23]

This demand for the balance of the two paths to knowledge and the conviction of their possible harmony not only pervades Harvey's work (E. Lesky has collected the corresponding passages), it also determines the practice of his research in the anatomy of the circulation as well as in embryology.[24] The widespread disparagement of Harvey's pronouncements in philosophy of science as the self-misunderstanding of an already basically "modern scientist" can be supported only if one takes his experimental research out of context and understands it as in itself complete science. In this way, Harvey appears, ex post facto, as already a "normal scientist" within the modern paradigm who best fulfils his task—the collection of facts and the testing of hypotheses—when he himself does not reflect on the paradigm.[25] But Harvey did not live in a time of "normal science"; and, as Kuhn writes, "it is . . . particularly in periods of acknowledged crisis, that scientists have turned to philosophical analysis as a device for unlocking the riddles of their field."[26]

We can draw still further consequences from what we have said so far. If for Harvey what is perceptually given is the "more manifest," "clearer, more perfect" (DG Pf, W.157); if it implicitly contains the universal, namely its appropriate principles, within itself, and is indeed the touchstone of the principles, then there cannot be, conversely, a *falsification of perception* through the universal. ". . . sensible things are of themselves and antecedent; things of intellect, however, are consequent, and arise from the former." A natural philosopher without full perceptual experience "would not better understand me than could one born blind appreciate the nature and difference of colours, or one deaf from birth judge of sounds."[27] Here, too, Harvey agrees with Aristotle: "It is also evident that if some sense experience is lacking, some scientific knowledge is necessarily lacking as well" (AP I 18, 81a38). For Harvey, there is no difference between "primary" (geometrical) and "secondary" sensible qualities, as Galileo, for example, projects it. Harvey, too, speaks of the "book of nature," but it is not, as with Galileo, "written in the mathematical language" with "triangles, circles and other geometrical figures," which we must first learn in order to "understand even a single word"[28]; rather, for Harvey, this book lies ". . . so open and so easy of consultation" before us, open to all the senses.[29]

Aristotle investigates the particular kinds of sensory perception in detail in the *De Anima* and says:

> By "proper" [object of a sense], I mean that which cannot be perceived by any other sense. . . . Every sense judges about these things and cannot be mistaken about the fact, for example, that something is a color or a sound, but only as to what and where the colored object is and what and where the object making the sound is (DA II,6, 418a12ff).

> For the perception of proper objects is always true; it belongs to all animals, but thought can be false, and belongs to no animal that lacks reason" (DA, III, 3, 427b12ff).

For Harvey, this truth of all the senses is still obvious and undeniable. Science needs our total attention, the undivided "physical" presence of the observer:

> And here, the example of astronomy is by no means to be followed, in which from mere appear-

ances or phenomena that which is in fact and the reason wherefore it is so, are investigated.³⁰

In astronomy, precisely this physical presence is impossible; that is what makes it, as Harvey says, so "uncertain and conjectural" (*tam incerta et conjecturalis*, ER 155; W. 132) All it can do is to invent systems that "save the phenomena," that is, that are merely consistent with them (*neque satis factum opinantur [sicut in astronomia] nova systemata ordinare, nisi omnia phoenomena solvant*; " . . . and there are persons who will not be content to take up with a new system unless it explains everything, as in astronomy" (ER 144, W. 123). A person who really wanted to investigate the causes of an eclipse, Harvey believes, "must be placed beyond the moon" (W. 124); he would have to be able, so to speak, to circumnavigate his phenomenon in order to grasp it with his senses and not merely through reason (ER 145f.; W. 124).

Thus if science rests on principles, these should not, however, be constructed by reason, but must emerge as *inherent* principles from the thing itself and its observation (*ex quibus in rei natura insitis, desumat argumenta*, ER 155, Willis, 132; cf. AP I, 2, 72a6f.) These are not simply universal foundations or "laws of nature," but the *peculiar principles* of the natural entities in question. They provide the possibility of certain knowledge because they specify the real causes of things and thus permit the recognition of their necessity: "The value of the universal is that it exhibits the causes" (AP I, 31, 88a 5), and "Perfect science of every kind depends on a knowledge of [all the] causes" (*scientia quaelibet perfecta, ex causarum omnium cognitione dependet*; DG 188; W. 362; W. omits "omnium"). How do we arrive at these causes? Harvey reflects about this in the preface to the *De Generatione* for the case of generation and development. We must trace back nature's steps from the mature organism (which is prior for us!) until we arrive at its "starting point," that is, the material and efficient cause (DG Pf). But the *goal* of the development is once more contained in this starting point as its true cause. In Chapter 50, which investigates the "fruitful" efficient causes in general (e.g. rooster, hen, egg), Harvey writes accordingly:

> . . . I call that fruitful which, unless impeded by some external cause, attains by its inherent force to its destined end, and brings about the consequence for the sake of which it is ordained.³¹

He puts it even more clearly in the *De Conceptione* (appended to the *De Generatione*):

> ... the 'final cause', both in nature and art, is primary to all other causes. ... There inheres, in some way or other, in every 'efficient cause' a *ratio finis* (a final cause), and by this the efficient ... is moved.[32]

The principle of the goal-directedness of natural processes, the idea of a global teleological order in nature, permeates Harvey's entire work. Included under this general principle is the priority of the organism as a whole to its parts or the primacy of function as against structure;[33] or the conception of the goal-directedness of all living motion (DML pp. 32f., 36); and finally, the axioms of Aristotelian physics (nature does nothing in vain, or without an end, nature does everything in the simplest way, etc.), which are to be found almost everywhere in Harvey's writings. Pagel writes:

> Harvey's most intimate concern would seem to lie in the question of purpose: what are the ends which Nature pursues in making blood circulate or causing individuals to take part in a seemingly eternal cycle of generation ...[34]

Only when it can give an answer to the question "What for?" is the science of nature complete; for Harvey, only finalistic principles represent ultimate explanations. That does not mean that—as in large parts of the *De Motu Cordis*—he could not also disregard teleology. In the *De Generatione* as well, the final cause has significance rather not so much in itself as in so far as it stamps the character of the other causes and thus serves to orient us in the search for the material and efficient causes; these, however, take center stage. It seems that the finality of nature does not always pose the decisive *questions* for Harvey, but that it does provide the obvious *framework* within which his investigations move, and without which they would be unthinkable.

A last question arises in this connection: what, for Harvey, is the goal of science itself? For this question, too, a passage in the "De Generatione" gives us a hint, namely the comparison of science and art:

> On the same terms, therefore, as art is attained to, is all knowledge and science acquired; for, as art is a habit with reference to things to be done, so is science a habit in respect of things to be known.[35]

Science and art are similar in their method; both rest, as Harvey explains, on experience and generalization, deduction and application.

But only art carries out the deductive path from a general-abstract representation to *actual production*. The goal of science, on the other hand, remains in the last analysis assured knowledge, which the principles grant in deduction—*and thus not the making or mastering, but only the pure knowledg of reality*. In this respect, too, Harvey's understanding of science is still stamped by antiquity.

Can we extract from this "science of the living" a guideline for the further investigation of the vital aspect in Harvey? It seems obvious that we should take the path indicated by him in the *De Generatione*: from the complete, but unanalyzed complexity of the phenomenon to its principles. While the Harvey literature has concentrated above all on retracing the path to the discovery of the circulation, we intend to take that as our starting point, as the "better known for us," that is, to investigate first the motion of the heart and blood in the *De Motu Cordis*. The *De Motu Locali Animalium* then treats in more general terms the phenomenon of living movement. This manuscript, published for the first time in 1959, has remained almost unnoticed in the literature; it is considered "unoriginal" and "scholastic."[36] However, it seems questionable whether originality should be the sole criterion for a work, which can at any rate provide insights into the thought of a scientist; if comment, reflection, and judgment on traditional ideas already disqualify a work, most of medieval literature must be considered uninteresting. We shall see whether the manuscript written by Harvey at almost the same time as the *De Motu Cordis*, considered as background for the movement of the circulation, cannot give us some sidelights on its vital aspect. Finally, our investigation will issue in the analysis of the movement that is "prior by nature," namely, the *development* of heart and blood as Harvey interprets it in the *De Generatione*; it will then move to the question of the origin of this development—of the being that is the ultimate foundation of the coming to be and the motion of heart and blood.

II. *DE MOTU CORDIS*: THE MOTION OF HEART AND BLOOD

<u>Summary</u>. *This chapter exposes, beneath what initially appears to be the mechanistic presentation of the circulation in the* De Motu Cordis, *a vitalistic substructure, which emerges as the genuine foundation of Harvey's physiology. (a) Over against the violent motion of the arterial blood through the heart there stands an inherent centripetal motion of the blood, which is manifested especially in the venous part of the circulation; the mechanical impulse of the heart serves to*

overcome the tendency of this "natural" motion. (b) This impulse in its turn, as we find further, represents a transformation of the original embryonic self-motion of the blood; the muscular mechanics thus has only the function of a secondary "impulse amplification." The mechanical aspect is integrated into the vital processes of nature through the basic principle of triggering and can at most be investigated in isolation for methodological purposes. (c) Finally, the circulation as a whole is not simply a geometrically and physically definable movement, but, as movement, exhibits a qualitative-vital aspect, which at the same time communicates its function in the organism. That is to say, the central motion of the heart serves not only for the distribution, but immediately for the revitalization of the blood, which transforms external motion into inner heat, calor innatus. *Thus the circulation is at the same time a cycle of regeneration for the vital heat of the organism; "efficient" and "final" cause collapse into one.*

The reduction of physiological processes to the vital level and their centralization under the governance of heart and circulation make it possible for Harvey to understand the vital aspect as the foundation for the autonomy of the living being over against its environment; from this point of view, the mechanical aspect emerges as an instrument subordinated to the goals of life.

Harvey's path to the discovery of the circulation has been investigated in all its detail by historians; in order to advance the discussion, we shall recapitulate here only the essential points of view as spoken of and in part controversially debated in the literature. There is no question but that Harvey relied on the results of Vesalius, Servetus, Colombo, Fabricius, and others—this prehistory of the discovery has been reported in detail.[37] Similarly, the role of Harvey's own observations, his rich experience in dissection and experimentation has often been given its due; without the knowledge stemming from embryological research and animal experiments, to which Harvey constantly refers in the *De Motu Cordis*, his discovery is indeed inconceivable.[38]

No longer uncontroversial, however, is the question of what kind of *conceptualizations* Harvey placed on this foundation. Harvey's calculation of the daily quantity of blood ejected by the heart was often seen as the core of the discovery even as the beginning of quantitative biology and medicine as such.[39] Against this, the objection was raised that Harvey's quantitative *demonstration* in Chapter 9 of the *De Motu Cordis* is not to be identified with the *path* to the discovery; that Harvey specifies only 1/36 of the amount now regarded as the minimum and that this "calculation" must rather be seen as a thought experiment in the context of his *qualitative* reflections.[40] Essential to these arguments, then, are considerations of *symmetry and analogy* (as between the lesser and the greater circulation, or between the right and left heart)—considerations that Harvey also stresses in the *De Motu Cordis* (DMC 15, W.14; 56, W.46); the idea of the identity of arterial and venous blood;[41] and finally, *teleological* aspects like the question of the goal of the venous valves discovered by Fabricius—these, as Harvey himself announced in a conversation with Robert Boyle, gave him the decisive indication of the way he should go.[42]

It is Pagel above all who has again made us aware of the element that was lacking and has investigated it thoroughly: namely, Harvey's symbol-imbued thought or what Ludwik Fleck would have called his "pre-ideas": basic images and ideas of a culture, which are handed down in certain concepts and are given ever new concrete expression and which, more or less consciously, guide the intuition and the knowledge of the investigator.[43] In Harvey's case, what is pertinent in this connection is the Aristotelian idea of the heart as "ruler" among the organs, which he expresses in his famous eulogy of the heart as the sun of the microcosm (DMC ch. 8). It is only as ruler of the circulation that the heart becomes the principal organ of the body ahead of the liver and the brain. Above all, it is the *symbolism of the circle* that shapes Harvey's thought and his writings. The circle, which in ancient times was the image of eternal celestial motion, became, in the neo-Platonic thought of the Renaissance, a universal principle, which Giordano Bruno, for example, or alchemical thinkers like Robert Fludd had already brought to bear on the heart and blood even if only symbolically. As Pagel aptly remarks, "Harvey's discovery and much of his observational work in embryology can be seen as the scientific complement and proof of these contemporary ideas."[44]

Naturally, the significance of these pre-ideas for Harvey has not remained undisputed either. Some people have spoken of analogies drawn after the fact and have objected that from the same conceptual background others too could have come to the same conclusions.[45] However, this objection—why precisely Harvey?—can equally well be raised against all other points of view. None of these was available to Harvey alone or on its own explains his discovery. The discussion about this priority in the literature, conducted chiefly along the lines: "It was X and not Y who helped Harvey to his discovery," seems idle. What is special in Harvey lies precisely in the broad scope of his abilities and interests and in the *synthesis* of different aspects. Induction in the Aristotelian sense is in fact more than the mere enumeration and generalization of data such as Bacon proposed in his *Novum Organon*. It represents an integrating, intuitive vision of prior experience and previous ideas brought together in a common, ordering principle.

The *De Motu Cordis*, to which we now turn, begins with a penetrating criticism of Galenic physiology, disavowing almost all its essential elements. Heart and arteries have no respiratory function; air cannot reach the heart nor blood pass from it to the lung (DMC 8f., W. 9f., 16 f., W. 15); nor do the arteries hold anything but blood (11, W. 11). Even if the arterial blood contains more spirits, it is still blood; for the spirits are not separable from the blood, but at one with it (*sanguis et spiritus unum corpus constituant;* 12, W. 12). Therefore, the arterial blood is the same as the venous (12, W. 11f); arteries and veins do not differ basically from one another just as the right and left chambers of the heart are structurally and functionally equivalent (14f., W. 13f.). The duality of the blood systems is thus cancelled just as their union through the pores of the cardiac system is denied (19; W. 17). Finally, Harvey indicates a principle of his physiological thought which radically contradicts Galen: the exclusion of attractive forces, thus of the *attractio* that had hitherto been a chief mover of physiological processes. Neither the arteries nor the heart move the blood through suction, that is, through active dilation; for expansion, Harvey says, occurs only passively and not autonomously, unless it happens through previous compression, as in the case of a sponge (13; W. 12).

In all this, a chief argument in Harvey's criticism refers to the "overloading of the phenomena" in the Galenic system. It has to do, for example, with the respiratory and relaxing function of the arteries: how is soot to be evacuated without the spirits' escaping (10; W. 10)? Or it concerns the structures in the region of the heart: why should the pulmonary vein have three or four different functions? "Nature is not wont to institute but one vessel, to contrive but one way for such contrary motions and purposes . . . " (*tam contrariis motibus et usibus unum vas et unam viam fabricare Natura solita non est;* 17, W. 16).

Thus, even in the introduction, not only does the fall of the Galenic paradigm become evident—even though Harvey has never mentioned it—but the new simplifying and centralizing tendency in Harvey's thought also becomes apparent. Particularly in connection with the exclusion of attraction, this could give the impression that Harvey sees in the system of the heart and circulation a structure governed solely by mechanical principles; only a closer examination corrects this impression. We shall now first investigate more closely the motion of the blood as the "primary phenomenon," and then, a step further back, the *heart* as the moving organ, before we turn to the *function* of the circulation in the *De Motu Cordis*.

a. THE MOTION OF THE BLOOD

To begin with, the circulation of the blood by no means appears in the *De Motu Cordis* as the quiet and "natural" circular motion that, according to the ancient conception, belonged to the heavenly bodies. Instead, it has a violent character, which Harvey emphasized in contradistinction to the comfortable flow of humors in the old physiology: the blood is driven through the vessels *vi et impulsu cordis, ictu cordis, impetu et violentia* (70, W. 57; 71, W. 58; 101, W. 84). Harvey illustrates the beat-like spread of the cardiac impulse throughout the arterial blood to a glove whose fingers momentarily rise up when it is blown into: "for in a plenum . . . the stroke and the motion occur at both extremities at the same time" (*in pleno . . . ictus, et motus simul sunt in utroque extremo*; 29f., W. 25). The *simultaneous effect at a distance* throughout the entire body of a motive force proceeding from the center, rather than local, attracting and regulating forces—that was the mechanical aspect of Harvey's revolution in physiology.

Nevertheless, when we look more closely, it appears that this power of the heart is limited to the arterial side of the circulation. Only here is there repeated talk of *vis*, *impulsus*, and *impetus*; only from arteries that have been opened does the blood spurt out violently, not from the veins (61 f., W. 51). In the periphery, the impulse provided to the blood by the heart is gradually lost;[46] "for it seems likely that the blood would be disposed to flow with sufficient slowness of its own accord as it would have to pass from larger into continually smaller vessels, being separated from the mass and fountain head." (*tarde enim satis* sua sponte *e majoribus in minores ramulos intrudi*, e massa et fonte separari; 77f., W. 63f.; author's emphasis).

Then how does Harvey explain the return flow of the veins? To the opposing arterial motion of the blood *sua sponte*, there corresponds an *inherent tendency to the center*, which combines with a "vis a tergo" that no longer proceeds from the heart, namely, the muscular compression of the veins (*declinante sponte sanguine, et venarum motu, compresso ad centrum*; 32, W. 27). This "willfulness" and tendency to collect characteristic of the blood in fact becomes the reason for the existence of the heart itself:

> . . . both because the blood is disposed from slight causes, such as cold, alarm, horror, and the like, to collect in its source, to concentrate like parts to a whole, or the drops of water spilt upon a table to the mass of liquid; and then because it is forced from the capillary veins into the smaller ramifications and from those into the larger trunks by the

> motion of the extremities and the compression of the muscles generally. The blood is thus more disposed to move from the circumference to the centre than in the opposite direction . . . ; whence that it may leave its source and enter more confined and colder channels, and flow against the direction to which it spontaneously inclines, the blood requires both force and an impelling power. Now such is the heart. . .[47]

The motion of the blood through the heart is in fact a violent motion. But in accordance with Aristotelian physics, it makes sense to speak of a "violent" motion only when a *"natural" motion is presupposed by and opposed to it*. Only because the stone actively seeks the center of the earth, does it present resistance to the thrower.[48] The blood is for Harvey not an "inert mass" in the sense of Newtonian physics, but a "mass with a will of its own," with a *centripetal tendency* comparable to the surface tension of water, a tendency that the heart is needed to overcome. It can impress a contrary impulse on the blood only for a time (the impetus is even called *virtus impressa*). So, the mechanical aspect of the circulation in Harvey rests, for one thing, on a pre-modern physics: it is not "action=reaction" that governs the occurrence, but active tendencies inherent in bodies.[49] Further, this physical aspect is *modified and serves only as a substructure* within the organism: the blood does not tend, like any arbitrary fluid, to the center of the world, but to the center of the *body*, the heart.

In another passage in the *De Motu Cordis* as well, the function of the heart is grounded in the unique tendency of the blood: already in the first stages of life, Harvey tells us, the blood shows "a tendency to motion and to be impelled hither and thither, the end for which the heart appears to be made." (*sanguis . . . moveri atque huc iluc impelli desideret, cuius causa cor factum fuisse videretur*, 196, W. 74). This appears to contradict what was previously said: here the tendency of the blood goes not to the center, but "back and forth", and the heart is evidently present not to overcome the motion of the blood, but *for its sake*. We shall see later on how this contradiction is resolved.

b. THE MOTION OF THE HEART

1. Basis of the Motion. If the motion of the blood in the *De Motu Cordis* cannot be accommodated within a purely mechanical perspective, this might, on the contrary, be possible for the moving organ, which, after all, Harvey himself did compare to a pump (see below).

In the second chapter, Harvey first describes the action of the heart in clear contrast to Galen; the heart moves like a muscle; systole, hence contraction, is its active movement, which leads to the pushing or pressing out of the blood (*protudere, exprimere*; 26, W. 22). Diastole, on the other hand, does not denote any attraction, but only a "reception" of the blood (*recipere*; 28, W. 23).

This is explained in more detail in the fourth chapter. The heart (that is, the ventricles) contains the blood from the pulsing auricles (32; W. 27); its contraction follows theirs: ". . . there is a short pause between these two motions, so that the heart aroused, as it were, appears to respond to the motion. . . ."[50] This becomes even clearer in the dying heart of an animal: it no longer responds to every action of the auricles, but can be aroused only with difficulty at every second or third beat until, finally, only the auricles continue beating.[51]

But if the ventricles are not the ultimate origin of the circulatory movement, but depend on the action of the auricles, the question then arises whether the auricles too are not "responding" to another movement, namely, that of the blood, which "is tending toward . . . the centre" and is filling them, as is said a little later (33; W.27). Harvey does not give us a direct answer to the question in this passage, but, instead, pushes forward to the last moments of life. Even after the cessation of the auricular pulse, Harvey says, ". . . an obscure motion, an undulation or palpitation, remained in the blood itself, which was contained in the right auricle, this being apparent so long as it was imbued with heat and spirit."[52] Thus the presuppositions of this *elementary spontaneous motion* are *calor* and *spiritus*; so long as they remain, Harvey observes, cardiac action can often be set going again even through the application of external heat.

However, the complement to this is to be found at the beginning of embryogenesis: in the fertilized hen's egg, the very first thing to be seen is a pulsing drop of blood (*gutta sanguinis quae palpitat*; 34, W. 28). In its systole, it almost disappears, so that "between the visible and invisible, between being and non-being, it produced its pulse and the principle of life" (*ita ut, inter ipsum videri et non videri, quasi inter esse et non esse, palpitationem et vitae principium ageret*; 36, our translation; cf. W. 31). The auricles are then formed from it and continue its pulsation, and with progressive development of the body, the ventricles finally appear.[53]

The investigation of the motion of the heart led Harvey, not to mechanical problems, but directly to the question of its ontogenetic origin in the *punctum saliens*. That this origin is not only a passing stage, but the latently enduring foundation of this movement becomes evident at the end of life. Harvey refers here to the Aristotelian *primum vivens ultimum moriens*: in the end, nature follows the path of its initial development in the opposite direction and returns to its starting point. *The principle, the "arché", of the organism appears at the beginning and at the end in its pure, original form.*[54] To be sure, Harvey modifies this Aristotelian thought: not the heart, but rather its auricles, indeed, strictly speaking, *the blood itself* is the first and fundamental living thing. Its original motion transfers itself to the structures developed and stabilized from it (*non cor solum esse primum vivens et ultimum moriens . . . sed auriculas . . . imo an prius adhuc ipse sanguis . . .*; 34, W. 29).

Thus the differentiated cardiac motion simply represents the *transformation of the primary motion of the blood*. In the same chapter, Harvey explains the meaning of this transformation through comparative biology. As embryogenesis begins with a pulsating drop or bubble of blood, so lower animals (insects, mollusks) have only a primitive heart bladder; in contrast to higher forms, they remain at this stage (*ita quibusdam animalibus [quasi ulteriorem perfectionem non adipiscentibus] pulsans vesicula . . . duntaxat inest*; 36, W. 30). This is further elaborated in Chapter 17: the simplest animals need no heart at all in virtue of their minute proportions; they use the contraction of the body itself to distribute nourishment (*quibus impulsore non opus sit quo alimentum in extrema deferatur*; 91, W. 75f). Other animals already have a heart bladder; larger and hotter ones need a ventricle as well for the longer stretches of the circulation. Finally, the most perfect animals have lungs and two ventricles.

In this way, moreover, we resolve the contradiction mentioned in the previous section between the direction of the motion of the blood at the beginning of life and that of the developed organism. The heart, as a sort of "impulse amplifier," assimilates the initial pulsation, the "movement to and fro" of the blood; but it then enlarges its amplitude for the sake of the higher development of the organism, indeed, to such an extent that a *centripetal tension* appears in the blood, which must now be overcome by the heart.[55] With the increasing size of the organism, therefore, the "natural" spontaneous movement of the blood experiences a "violent" dilation and acceleration; in other words, *the "vital" aspect of the circulation itself produces the "mechanical" aspect*.

Harvey goes back to an even earlier stage in this remarkable fourth chapter on cardiac activity. We must ask whether the pulsing blood is to be understood as the beginning of the movement of life, or not rather the rhythmic action of conception itself. "The semen of all animals—the prolific spirit . . . leaves their body with a bound and like a living thing" (*quandoquidem et sperma animalium omnium . . . et spir-*

itus prolificus palpitando exit, velut animal quoddam; 34, W. 29). Thus in the first chapters of the *De Motu Cordis*, starting from cardiac motion, conception comes into view as its analogue and at the same time its ultimate origin.

2. Mechanism of the Motion. The fifth chapter appears to turn at last to the mechanical view of cardiac motion for it compares that motion to a machine with interlocking gears and with the mechanism of a gun lock, where the shot takes place via a series of consequences resulting from activation of the trigger (*nec alia ratione fit, quam cum in machinis, una rota alia movente, omnes simul movere videantur; et in mechanico illo artificio quod sclopetis* (=arquebus or carbine) *adaptant . . . et sq.;* 37, W. 30). The analogy is the only one of this kind in the De Motu Cordis; we shall examine it somewhat more closely.[56]

To begin with, one is struck by what it is that Harvey wants to illustrate with this analogy, namely, the harmonious arrangement of the actions of auricle and ventricle. In their movement, "there is a kind of harmony or rhythm preserved between them, the two concurring in such wise that but one motion is apparent" (*servata quasi harmonia et rhytmo, ut ambo simul fiant, unicus tantum motum appareat*; ibid.). Since the theme "harmony and rhythm" also occupies a good deal of space in the *De Motu Locali Animalium*, we shall take account here of two passages in that manuscript; this seems all the more justified since Harvey himself, in a later section on cardiac motion (DMC 97f, W. 81) refers to that work, which he had conceived at about the same time as the *De Motu Cordis*.

Both passages have to do with the role of the mechanical in nature. First, Chapter 20, *De Artificio Mechanico Musculorum:*

> Mechanics: whatever controls that through which nature is overcome, and helps with the difficulties [that arise] when something unnatural is useful.[57]

"Mechanics," then, is the "outwitting" of nature for human ends—but can there be something mechanical *within* nature? Harvey continues:

> Thus in the muscles, nature, too, never stops helping with such problems. Thus here too there is much to be marveled at, how muscular forces move bones and the weights attached to them.[58]

The paradox of a "self-outwitting" of nature is thus solved, in that nature, in the sense of the living makes use of mechanics in order to overcome the inanimate and the heavy, and integrate it into living movement. *Mechanism as the "suspension" of the purely physical becomes the "artifice" of living nature itself (artificium mechanicum)*. But the heart, too, presents just such an artful arrangement of muscles and mechanical structures.

The second passage (ch. 14, p. 96f) quotes at length from the *De Motu Animalium*, where Aristotle at first compares living movement with that of automata, which are set going by a release mechanism (like the release of a string). But, as Aristotle continues:

> ... in automata ... no alteration [*alloíosis*, lat. *alteratio*] takes place in this process ... ; in the organism, on the contrary, the same organ can grow larger or smaller and change its shape since the parts increase through heat and in turn contract through cold and are thus altered.[59]

Thus, while in automata, only unalterable parts affect one another from without, living movement proceeds from the inner, qualitative change of definite parts, in order also, admittedly, to include mechanical effects on rigid body parts (bones, among others). Harvey localizes that alteration in the muscles and in the heart.[60] But Aristotle, too, appears to speak in the same passage of the heart as the "origin" of movement, where ". . . a small change . . . causes many large differences at a distance. . . ." (701b25–26). This thought, too, Harvey takes up later:

> Nature moves by signs and watchwords, made with small momentum in the area of the principal parts (Aristotle). It is as with shooting: a small momentum in the beginning, a great error at the end.[61]

Here we return to our initial analogy in the *De Motu Cordis*: the triggering of a gunshot. "Small cause, great effect": one could encapsulate, in this way, the mechanical principle that animate nature applies. It is realized through a lever- or cascade-like arrangement of spontaneously moving and rigid parts, which is activated "in response to a sign" (*tanquam signo dato*, DML, p. 146). Thus, we see what is important for Harvey in his mechanical analogies for cardiac motion: they illustrate the interaction of the rhythmic movements of the parts to produce a *harmonious totality of motion, which serves for the mechanical strengthening of spontaneous living movement*. But that this moving whole exists as such even before its later analysis into individual components is demonstrated in embryogenesis by the initial pulsation of the blood from which the pulsating parts are to be formed. *The blood, the later object of motion, is thus ontogenetically "earlier," its subject.*

Even in this passage in chapter five of the *De Motu Cordis*, however, we still find no answer to the question of the release mechanism, the "watchword" to which the auricles respond. Still, we can surmise for the moment that the blood does not become the mere object of cardiac motion, but instead, itself remains the enduring foundation of that motion, in that it is actually the blood that triggers the motion through its streaming into the auricle.[62]

Nevertheless, in the developed circulation the activity of the heart undoubtedly outweighs that of the blood. Whether, Harvey continues, the heart, in this activity, also transmits to the blood something more than movement, namely, *calor*, *spiritus*, or *perfectio*, would need further investigation (DMC 38, W. 32). It is to the question of the "final cause" of the motion of heart and blood that we now turn (57, W. 47).

C. THE FUNCTION OF THE CIRCULATION

1. Blood and Heat. Harvey gives an initial answer to this question in the eighth chapter: the heart possesses not only the beat with which it moves the blood, but also a *virtus*, which perfects the blood, nourishes it, and prevents corruption or congestion. The blood is in fact threatened by this when it flows through the cooler periphery; only in the heart does it recover *calor* and *spiritus* (57, W. 47). Thus, the central principles of Galenic physiology crop up again here, if only, as will appear, in essentially altered form.

This is explained more precisely in chapter fifteen. Here, invoking Aristotle, Harvey explains heat as the principle of life as such; death is corruption (*corruptio*) through the loss of *innate heat* (82, W. 68). Therefore, the heart is needed as a source of heat, which must be located in the center of the body in order to insure an even distribution.[63] On the one hand, heat controls all vegetative processes (*concoctio, nutritio et omnis vegetatio*; 82, W. 69) so that the real digestion of food actually takes place not in the liver, but in the heart (84, W. 70). On the other hand, the vitality of the body parts is also bound to heat; *motion* is inconceivable without it for cold means stiffness (83, W. 69). There is no mention in this passage of a third task assigned to heat by Aristotle, that is, making *sensation* possible (although it is mentioned in the *Praelectiones*, p. 310).

What is the character of the *calor innatus* in Harvey? Surely not that of a "fire"; if he calls the heart the "hearth" of the body, he does so in a metaphorical sense (*lares corporis, lar iste familiaris*; 57, W. 47). The Galenic conception of a cardiac fire, controlled through the well apportioned, tempered, and filtered supply of blood and air, is incompatible, not only with Harvey's altered anatomy, but above all with what is now the immense speed of the circulation.[64] Here again, Harvey calls on Aristotle, who rejects a simple adoption of the element of fire in the physiology of the organism (see note B 21). The *De Respiratione* describes the heat of the heart as a "foaming up," but not as combustion of the blood; the *pneuma* arises from its vaporization (*De Resp.* 20, 479b25ff, 480a2ff). In the *De Generatione Animalium* as well, Aristotle distinguishes from fire the heat inherent in the semen and the blood, and connects it to the *pneuma*, which no longer corresponds to a terrestrial element, but rather to the "primal stuff of the heavenly bodies," the ether (DGA II,3, 736b35f). Its essence, however, is *self-motion*, and indeed, *circular* self-motion (*De Caelo* I,2, 269a24ff). The ultimate material principle of the generation and the motion of living things can no longer be of the nature of the terrestrial elements.

In the *De Generatione*, Harvey confirms this non-physical, "divine" nature of the heat of blood and semen (DG 332, W. 505f). In the *De Motu Cordis*, it is above all the aspect of movement that is important; let us turn again to chapter 15:

> For we see motion generating and keeping up heat and spirits under all circumstances, and rest allowing them to escape and be dissipated (*motu enim in omnibus calorem et spiritus generari et conservari videmus, quiete evanescere*; 82, W. 69).

Thus, motion is able to generate and maintain the heat of the organism for this heat arises through the motion of the blood itself.[65] The blood was "required to have motion" in order not to "congeal" (*sanguini itaque motu opus est . . . immotus coagularetur, ibid.*). But in the "more confined and colder channels" of the periphery, its spontaneous motion is in danger of being exhausted (85, W. 71), which would be synonymous with its curdling, its "death" (cf. 82, W. 69). Hence, it needs the violent motion and impulsion provided by the heart in order to be sent out to the arteries hot, enlivened, and renewed, and to warm the organism (*a frigore extremorum . . . consistens aut gelatus sanguis, et spiritibus [uti in mortuis] destitutus*; 82, W. 69; *tum violentia opus habet sanguis, tum impulsore: quale cor solum est*; 85, W. 71).[66]

With his question in chapter five, whether the heart contributes anything to the blood other than motion, Harvey had initially distinguished between a kinematic and a qualitative aspect of cardiac action. Now both are again united: the *pulsus* of the heart is nothing but its own *virtus*: it is one and the same act of the heart ". . . by whose virtue and pulse, the blood is moved, perfected, made apt to nourish, and is preserved from corruption and coagulation" (*cuius virtute et pulsu sanguis movetur, perficitur, vegetatur et a corruptione et grumefactione vindicatur*; 57, W. 47). The mechanical aspect of the motion of heart and blood can indeed be *considered* separately from its qualitative=teleological aspect, but it cannot *exist* without it *since the one coincides with the other*.

For that reason, the arterial side of the circulation is characterized by heat and by violent mobility of the blood (57, W. 47); in a condition of fever, the pulse is stronger and more violent (86, W. 72); it forces itself further into the periphery and can be felt in the fingers (101, W. 84). Yet its force decreases with distance from the moving organ (77f, W. 63f), and in the cold periphery, the blood becomes cooler, stiffer and "exhausted" (*effoetus*; 57, W. 47). Thus, the circulation exhibits, along with its hydraulic component, a *qualitative gradient*; and the centripetal impulsion of the blood is nothing but its effort to maintain itself.

This regenerating cycle of warming and subsequent cooling or thickening imitates the continuous and self-sustaining circular motion of the fifth celestial element.[67] But two other "pre-ideas" also vouch for this: on the one hand, alchemical distillation, which was also called *circulatio*, and which had already been applied, if only qualitatively, to the motion of the blood by Andrea Cesalpino (1523–1603);[68] on the other hand, as mentioned by Harvey himself, the circulation of water, which like blood "emulates the circular motion of the superior bodies."[69]

As the rain fertilizes and animates the earth, so the circulation serves not only the maintenance of the blood itself, but also the nourishment and warming of the tissues (56, W. 46). The relation *motus* > *calor* is reversed in the periphery. Here, it is heat that first makes motion possible; without it, the limbs become stiff and immobile (83, W. 69). Thus, the final cause of the circulation is of a double nature: maintenance of the blood through the heart, maintenance of the tissues and the limbs through the blood. Both coalesce, however; for even the heart needs the coronary vessels to provide its own heat as a prerequisite for its motion.[70] *Thus the qualitative-vital aspect of the motion of the heart and blood also closes in a circle*: heat makes possible the motion of the heart, this in turn produces and maintains heat in the blood: *calor > motus > calor*. The efficient and the final cause of the circulation are indissolubly bound with one another in the heat of the moving blood as principle of life.

It follows from the character of heat as described above that, unlike the Galenic fire, it does not need fuel (*pabulum*); neither the blood nor nutriment are considered as such in the *De Motu Cordis*. Nor does the air find any entrance into the blood in order to support cardiac heat; the function of the lung—here Harvey is cautious—plainly consists only in cooling and tempering.[71] Connected with this is a further fundamental reinterpretation: the spirits no longer arise as a special, subtle stuff from blood and respired air, but are *identical with the heat and animation of the blood itself*; Harvey now uses *calor* and *spiritus* as synonyms.[72]

The vital heat does not arise from a material process of transformation, nor does it need anything added from outside. Even if the blood takes nourishment into itself and gives it up again in the periphery (84, W. 70), the "energetic" side of the circulation remains untouched by this; through the coincidence of efficient and final cause in heat, the circulation becomes a *closed system* or, like the eternal self-motion of the heavenly bodies, a "perpetuum mobile."[73]

Harvey is not thinking here in the categories of modern physics or even about the abrogation of physical conservation laws. It is rather the criterion of living things that they carry within themselves their own source of motion, independent of external processes—that they move themselves in the sense of growth and metabolism as well as in the sense of locomotion. The circulation as a spatial and, at the same time, qualitative cycle is the physiological foundation of this self-motion. *Thus, the autonomy of the living rests in the last analysis on a specifically vital property of the blood, that of being warm "in virtue of its essence"*, that is, of transforming the (local) motion of the heart into the inner, qualitative movement of vital heat—which, in its turn, makes possible the self-motion of the parts, including the heart.

2. Tendencies of Harvey's Physiology. It would certainly be inappropriate to consider the *De Motu Cordis* as a treatise propounding a new physiology. Nevertheless, where Harvey goes beyond anatomical demonstration in laying the foundations of his theory, the direction of his physiological thought also becomes apparent in many ways. Starting from the Galenic system, we can describe it with the concepts of *reduction, centralization,* and *autonomy*.

To be sure, Harvey was thus following general tendencies in the thought of his time; yet it was also the impetus of his discovery itself that prescribed this direction to him. The violence and the dynamic of the flow of blood also poured like a flood through the edifice of the old physiology and tore out with it not only furniture, but whole floors of the building. With the circulation as the unitary substrate of bodily processes, the boundaries between the different compartments of the body—the arterial and venous systems; the head, thoracic, and abdominal regions—became permeable or wholly invalid. Not only that: till then, physiological processes had followed ordinary stretches of time (Riolan, in his last attempt to save the Galenic paradigm, wanted to allow the blood to circulate, at most, once in twelve to fifteen hours).[74] But now, the whole mass of blood streamed through the heart and body with high velocity (*tota massa brevi tempore illinc pertranseat*; DMC 58, W. 48), stormy, uniform, and unceasing, so that no time seemed to remain for the complex local metabolic processes of Galenic physiology, its regulation through attraction, repulsion, and so on. "Mixing and confusion of the humors"—that was not for nothing a chief objection of Riolan to Harvey's theory.[75] Bylebyl has shown that many processes that were pathological according to the Galenic conception (for example, the excess of blood in inflammation, the violent expulsion of the blood from the vein in blood-letting) become normal conditions in Harvey's system.[76] Thus, all the traditional interpretations of physiological processes had to be considered anew, and Harvey did this necessarily in the light of the one thing that was now certain: the circulation with its unifying dynamic.

Reduction: We find in the *De Motu Cordis* only two remaining levels of explanation: the anatomical-hemodynamic level on the one hand, and its vital-qualitative basis on the other; in this way, Fernel's hierarchy has already been considerably reduced. Thus Harvey no longer recognizes any different kinds of spirits. They again become the unified vital medium of the pre-Galenic period, no longer as a special stuff, but as the vital aspect of the blood itself. It is even more striking that nowhere in the *De Motu Cordis* does Harvey mention the *soul*. However, as we shall see, this does not mean renouncing a final vital principle of the organism; but it does mean abandoning localizable "parts of the soul" with specific "faculties", especially the "vital faculty", which, in Fernel, governs the function of the heart and the pulse.[77]

Centralization: Against Galen's "multipolar" and "decentralized" physiology, Harvey sets the sovereignty of the heart, the "sun of the microcosm" (57, W. 47). It is not dependent on liver or brain, but possesses life, sensation, and motion even before those organs have arisen in the embryonic period. "... (L)ike a kind of internal creature," it forms and nourishes the body; on it, "all power depends in the animal body" (*tanquam animal quoddam internum . . . a quo . . . omnis in animale potestas derivetur*; DMC 100, W. 83). There is no talk in the *De Motu Cordis* of a blood- or spirit-forming function of the liver; its role consists only in the preparation of the chyle for its proper perfecting in the heart.

Local regulation of flow is replaced by central regulation: the velocity and quantity of the flowing blood vary according to inner and outer influences on the heart (muscular motion, nourishment, affects, and so on; 60f.; W. 50). The periphery receives only the blood and is filled with it.[78] A pathology determined primarily by the heart is also suggested: emotions influence pulse, heat, and constitution of the heart, and can give rise to incurable illnesses through a weakening of the central source of heat and nourishment (83f., W. 70). Poisonings and infections reach the heart, where the fate of the afflicted person is decided (85f., W. 71f.). Harvey also identifies a lack of blood or its overabundance as causes of death (66, W. 54) as well as the loss of the *calor innatus* (823, W. 68).

<u>Autonomy</u>: *The independence and self-motion of animal life begins with the pulsation and heat formation of blood and heart.* This is evident precisely in the lower animals, which swing back and forth between warmer phases of pulsation and motionless rigidity in the cold season of the year. They are "of a doubtful nature," occupying the boundary area between animal and plant life (*dubiae sunt naturae, . . . videantur . . . quandoque vitam animalis agere, quandoque plantae*; 91, W. 76). Only with the closed system of the blood does the organism become autonomous over against its environment.

According to the traditional conception, metabolic movement ran through the organism, as it were. Nutriment and air streamed in, in order to be used again in the periphery, as they were transformed into blood and spirits. In its permeability to this "vertical" movement of air and earth, animal life remained largely tied to the vegetative life of plants; this was reflected in the calm and circumspection of the organism's currents and metabolic processes in the Galenic image of the "terrace garden." Its central heat alone gave the animal a more than vegetative life. Only with the closed system of the blood, which no longer serves merely as a transitory stage, but maintains itself in its motion; and which, continuous and uniform, is in simultaneous movement as a whole, like the planetary spheres: only now does the organism face the environment with an anatomical substrate that gives it autonomy and its own dynamic. It becomes something *moved in itself*, and this is the presupposition for its external movement in place.

It must have fascinated a biologist like Harvey, who undertook comparative investigations in the whole animal and plant kingdoms, to discover in the circulation a key to the understanding of animal life. From this point of view, we can understand why he did not want to allow the plantlike-vegetative element any validity even where it does indeed have its place as in the capillary region of the periphery or in the hepatic portal vein system; nor did he ever recognize as such the chyle or lymph vessels discovered by Aselli in 1622. Harvey was inclined to absolutize the newly discovered autonomy of the circulation. That becomes clear above all in his "Heraclitean" conception of the spirits, which he identifies with the (innate) *heat* while he excludes the *air* from the blood. It is also evident, however, in the character of the blood as a uniform substrate, which does indeed go through a qualitative cycle with respect to its vital heat, but is scarcely modified in its *material* aspect. The chyle dissolves in the blood like a drop of water in a barrel of wine (87, W. 73); arterial and venous blood do not differ from one another in material composition (12, W. 11d; cf. also ER 132f., 113ff.).

All this leads to the conclusion that Harvey identifies the qualitative polarity of the circulation with the kinetic and considers the opposition of center and periphery as decisive while he entirely eliminates the qualitative difference between the arterial and venous systems as Galen had conceived them. This one-sidedness was later corrected—but only on an atomistic foundation (see below, Part D).

d. CONCLUSION

Our exposition so far may evoke some objections, and in particular this one: is not the *De Motu Cordis* really concerned with quite different matters? Is it not a question of hydraulic considerations, quantities of fluid, experiments on the relations of flow? And is not the central thesis of the treatise that the blood, expelled with great force by the heart, moves back to it in a circle?

All this is not to be denied. Nevertheless, on closer inspection, if we do not assume in advance that Harvey's apparent digressions on embryological and teleological questions are merely incidental, and if we examine the *De Motu Cordis* against the background of Harvey's other writings, then a substructure appears everywhere beneath the obvious "hydrodynamic" level—a substructure which does not indeed carry the argument, but does in fact represent the genuine physiological foundation for what is being demonstrated, and thus what is "prior in nature" in the phenomenon of the motion of the heart and blood. Only when we keep this substructure in view do we avoid understanding one-sidedly, under a modern aspect, the concepts that Harvey uses for the description of his phenomenon.

We then recognize that the "violent" motion of the blood presupposes a natural motion; that the principle of cardiac motion is anticipated embryologically in the elementary self-motion and pulsation of the blood, which is then transferred to the more elaborate structures that have developed; and that their mechanics represents only an instrument, a "trick" of nature itself, in order to effectuate the goal of a higher development through the amplification of an elementary self-motion. *The vital aspect, most purely visible in embryogenesis, itself produces the mechanical aspect and is overlaid by it.* But that the vital remains the foundation even in the developed organism is indicated in the *De Motu Cordis* in two respects. For one thing, cardiac action is comparable, not to clockwork, but to a *triggering* mechanism: but the triggering and the reaction to it cannot themselves be interpreted in mechanical terms. We shall see later how Harvey explains them. For another thing, cardiac action, like muscular contraction, is made possible only through the qualitative cycle, which rests on *the vital capacity of the blood for heat*. It transforms the kinetic into the qualitative action of the heart, that is to say, *pulsus* into *virtus*. The blood is able to generate heat and to communicate it to all the tissues and organs, including the heart, not through the fuel of a fire, but through its own movement. Thereon rests the foundation of the autonomy of the organism over against its environment.

That the efficient and final causes of the circulation converge in this way in the blood, and that the apparent object of motion is its true agent even in the mature organism will become even clearer later in our exposition. What is essential is indeed already provided in the *De Motu Cordis*; yet Harvey remains cautious there. Physiology had first to be rebuilt. To venture out into an uncertain area seemed dangerous; after all, one might fall back, unintentionally, into the rut of the Galenic paradigm or one might argue into the very hands of one's opponents. Thus, even later, Harvey always emphasized the fact that in the *De Motu Cordis* it was a question primarily of the presentation of the phenomenon, of the establishment of the "that" before that of the "why" (see below.) This in no way stands in contradiction to his conception of science. Aristotle says in the *Posterior Analytics*:

> ... we certainly cannot know the why before the that; plainly, in the same way, we cannot know the essential nature of something without knowing that it exists. For it is impossible to know what something is without knowing that it is. (AP II,8, 93a18ff.)

Nevertheless, it is also the case that:

> Knowledge of the that together with the why is more accurate and prior ... (I,27, 87a31f.)

Harvey finds himself in complete agreement with this when he writes:

> And first I own I am of opinion that our first duty is to inquire whether the thing [sc. the circulation] be or not, before asking wherefore it is? For from the facts and circumstances which meet us in the circulation admitted, established, the ends and objects of its institution are especially to be sought.[80]

It is scarcely credible how this passage could repeatedly be taken to show a preference for a science of plain, verifiable facts.[81] It is the inescapable path of our knowledge that it must advance from the better known, the phenomenon, to its principles; that Harvey did not yet consider this relationship established with certainty in the case of the circulation does not mean that scientific knowledge had to be satisfied with the mere fact! For only this relationship makes the fact intelligible as a *necessary* fact, namely, by specifying its real causes. Accordingly, although the reflections on the final cause of the circulation are admittedly presented in the fifteenth chapter of the *De Motu Cordis* as "probable (= not yet certainly established) reasons" (*rationes verisimiles*, 82; W. 68; our rendering), they do however specify why "the circulation is matter both of convenience and necessity" (*quod ... ita esse et conveniens sit et necessarium*). They are principles from which the circulation will be proved *deductively*, "a priori," in contrast to certain consequences, which the next chapter offers as "a posteriori" arguments (85, W. 71).

Harvey is fully aware of the fact that his treatise leaves open more questions than it answers. He raises them himself, suggests possible answers, and refers to future writings: to the *De Motu Locali* for the question of the nature of contractile motion (97f., W. 81); to the *De Generatione* for the question of the original organ of the body, of the vital principle, and of the self-motion of the blood (88f., W. 74). Only the solution of these problems places the circulation into the context of a full-fledged science. We shall turn to these writings in the coming chapters.

Thomas Kuhn writes in *The Structure of Scientific Revolutions*:

> ... discovering a new sort of phenomenon is necessarily a complex event, one which involves recognizing both *that* something is and what it is ... But if both observation and conceptualization, fact and assimilation to theory, are inseparably linked in discovery, then discovery is a process and must take time.[82]

True, Harvey would not have accepted this "inseparability" and would have insisted that the phenomenon he had discovered existed independently of its conceptual interpretation. But Kuhn might have been able to call his attention to one thing: the very establishment as a fact of the phenomenon discovered by Harvey is part of the *social* process of "discovery," which is no longer confined to a purely phenomenological level; even here, the new conceptual structure must be fought for and contested. Harvey's restraint in this connection, his insistence on the phenomenological interpretation of the *De Motu Cordis* betrays his modesty. Still, it encouraged the disregard of the vital aspect of his treatise and, subsequently, a chiefly mechanistic interpretation of the circulatory process.

III. *DE MOTU LOCALI ANIMALIUM*: THE MOVEMENT OF LIVING THINGS

<u>Summary</u>. *The work on the local motion of animals* (De Motu Locali Animalium, *1628), which has been almost entirely ignored, nevertheless contains essential points of departure for an independent Harveyan vitalism. Despite its ostensibly more general theme, it proves in actuality to be an important stage in Harvey's reflections on the cause of the rhythmic action of the heart and the circulatory movement of the blood.*

(a) Harvey proceeds from the traditional explanatory principles according to which calor innatus and spiritus play the role of mediators between soul and bodily movement. But in Harvey, movement becomes polar in nature: there is now antagonistic action, motion back and forth, expansion and contraction, and so on. Since he cannot reconcile this polarity with the traditional principles of explanation, he takes a new route. He fuses calor and spiritus, which till then represented

independent agents in the service of the soul, with the structures and tissues participating in movement, and their actions, too, he now interprets as related to one another in polar fashion.

(b,c) On the one hand, this polar relation needs the power of self-motion in the structures concerned, in particular, muscular contractibility; but on the other hand, it needs a kind of "perception" of the preceding action in each case, a perception which then issues in an "answer." This leads Harvey to the conception of an elementary "sensus" of organic tissues, a concept that lies at the basis of Glisson's and Haller's later doctrine of "irritability." Rhythm, for Harvey a central phenomenon in the movement of the organism, thus arises through reciprocal perception and reaction of the mobile structures. In this process, the nervous system is given only a modulating task; for Harvey, living movement cannot find its explanation in a central (apolar) directive principle, whether it be the soul or the brain.

Applying this conception to the circulation, Harvey sees the solution to the riddle of the rhythmic action of the heart in the reciprocal relation between the influx of blood and the cardiac reaction, that is, in the triggering of a contraction of the irritable cardiac musculature. In this way, a free rhythm arises, which can in its turn be reshaped and modulated by psychological factors.

In the *De Motu Cordis,* Harvey ascribes sensation and motion to the heart independently of the brain (DMC 100, W. 83). The *De Motu Locali* takes up this theme in a general way: *sensus* and *motus* are what distinguish the living from the non-living, the human being from the corpse (*animal ab inanimali, homo a cadavere difert [sic!] sensu et motu*; DML 14). Clearly, we may expect from this work further illumination of Harvey's view of the motion of heart and blood. Granted, the analysis involved is not altogether simple: Harvey's remarks, frequently almost like catchwords, are often unintelligible at first reading and need careful interpretation. In the last section of this chapter, the chief results will be supported by reference to later works.

After some general reflections, Harvey's investigation leads first to the differentiation of various sources of living movement. In first place stands the soul; sensation and desire (*sensus* and *appetitus*) with their respective objects follow; then come *spiritus* and *calor* as mediators of movement, and finally, the individual contractile or mobile parts like ligaments, sinews, muscles, and joints. A number of passages in the *De Motu Locali* point to the cooperation of these sources of motion. As the living thing considered as a whole is moved by the soul, in the same way it is moved in its parts by the spirit (*sicut animal movetus ita corpus a spiritu ita musculus*; p. 96). In the movement that the hand performs, for example, muscles, spirit, heat, and soul operate in pursuit of a common goal (*quodcumque vero manus prestat illud musculi per manum et spiritus per musculos et calor per spiritum et anima per corpus et communi huic*, p.139).

In what follows, we shall consider more closely the sources of movement that take center place for Harvey himself; *calor* and *spiritus* on the one hand and the muscles with their nerves on the other.

a. *CALOR* AND *SPIRITUS*

In the *De Motu Locali* as well as in the *De Motu Cordis,* heat is the necessary condition for movement and sensation: *sensus non sine calore, nec motus* (p. 92). Cold numbs and prevents mobility (*frigore torpet ut sensus ita motus*; p. 102). In the other direction, [muscular] movement generates heat and thus sensation for which reason the latter, too, proceeds in the last analysis from the beating heart (*caro utilis . . . caliditate ad sensum . . . ergo corde sensus*; p. 88). Nevertheless, in contrast to the *De Motu Cordis,* the spirits are not simply identified with heat, but are first discussed separately. It was only in his later works that Harvey transferred the characteristics of the spirits entirely to the blood on the one hand and to the muscles on the other.

Chapter 14, *De Spiritu Motivo,* refers to Aristotle, who in the *De Motu Animalium* introduces the *pneuma* into the physiology of movement as the "middle" between soul and limbs, that is, as an instrument which "contracts and expands without external force" and thus produces pulling or pushing movements in the organism (DMA 10, 703a5–28). Harvey, too, sees in spirit the principle of spontaneous self-motion (*principium eorum quae sponte oriuntur*, p. 94). Its capacity for expansion and contraction produces on the one hand involuntary movements like the pulsation of the heart and blood (namely, in the *punctum saliens* of the embryo) or the pulsing motion of ejaculation (*ibid.*). On the other hand, voluntary contraction of the muscles also derives from the movement of the spirits (*a spiritu caro ipsa contrahitur . . . et vita musculi in spiritu; ibid.*) Here, Harvey avoids speaking of an active expansion through the spirits: instead, relaxation occurs through the antagonistic muscle (*musculus contractus . . . ab antagonista relaxatur et sic vicissim*; *ibid.*).

Thus spirit is present in all movements for "the power of movement must be assigned to a unified organ" (*opportet ad aequale organum vim in omni motu assignare*; p.96). Again, increasingly larger, more solid, and more differentiated parts of the body serve only for the *mechanical strengthening of this elementary motion*:

> ... the more difficult a movement, so much the more inclusive and stronger is the organ (of movement) for mechanical reasons, for which reason many animals have only spirit, others in addition fibers, flesh, (sinews), and so on.[83]

Following Aristotle, Harvey now connects the movement of spirit with "alteration" (*alloíosis*). What is meant is a qualitative change in the tissues, which takes place through a change in temperature. According to Aristotle, it is the *pneuma* that transforms heat into expansion and cold into contraction of the tissues; thus the spirits should have the same function.[84] However, a difficulty arises here for Harvey. Experience shows that it is heat above all that makes possible power and movement (*magis videtus calor vigorare . . . quam frigus*; p. 102), but for Harvey, movement is primarily muscular *contraction* (*omnes motus a contractione, ibid.*). Is it not then the case that a connection seems to exist contrary to that specified by Aristotle, a connection, namely, between heat, spirit and *contraction*? And does not cold produce weakness or total immobility of a limb rather than its contraction (*frigore . . . impotentia movendi*; p. 102)?

The Aristotelian doctrine of *alloiosis* is connected above all with the *heart*, which is able to perceive the change of temperature and, as the source of movement, to communicate it to the body.[85] However, for Aristotle it is *diastole*, the expansion caused by the foaming up of the heated blood that is the principal action of the heart (*De Resp.* 20, 479b27f.). Harvey, in contrast, must instead bring the doctrine of heat into harmony with the active contraction of the heart in *systole*. Hence his doubts: does movement occur primarily as filling and expansion or rather as emptying and contraction?[86]

In the correlation of heat and dilation, cold and contraction, Harvey is unable to follow the Aristotelian doctrine of movement. On the other hand, heat by itself cannot now be equally responsible for both the contrary poles of living movement—for which reason Harvey speaks, for example, of the reciprocal relaxation, that is to say, stretching, of antagonistic muscles (see above). Here lies an essential reason for Harvey's reinterpretation of the doctrine of spirits. In what follows, it will emerge that he does indeed retain spirit and heat as a general prerequisite for organic movement, but nevertheless, transfers the reason for contraction or dilation to the peculiarity of the parts themselves. Thus, heat is a precondition for contraction in muscle, but also for dilation in the blood; and this polar distribution of the principle of spirit becomes the foundation of a *reciprocal relation* of the mobile parts.

At this point in his discussion, Harvey at first continues with his critical questions about the *spiritus* doctrine. Is the *spiritus movens* localized in the muscles or in the blood? Is its source the arteries or the nerves—in other words, the heart or the brain? If the latter, do motor and sensitive nerves then differ (as Galen taught), or are the motor nerves also sensitive? In that case, the *movement* proceeding from the brain would actually be identical with the *sensation* going to the brain (. . . *motus a cerebro idem cum sensu ad cerebrum*; p.100). Harvey himself presents the answer to the first pair of questions: the spirits proceed from the *heart* as the organ of movement; they flow through the arteries, but are one with the blood as well as with the muscles.[87] The exclusive sources of life in the parts are thus the motion of the heart and the ubiquitous blood; but then the question arises of the role of the nerves and brain.

In reply, Harvey points to two experiments. Compression of a nerve leads to insensibility of the limb with continued movement; and in a decapitated chicken, whose arteries have, however, been ligatured so that the movement of the blood is still preserved, uncoordinated, convulsive movement is exhibited (p.102f.). With the supply of blood and heat, Harvey concludes, there is already *an autonomous mobility of the parts that is not bound to the nerves*; only purposive, goal-directed movement needs the nerves, insofar as these in fact mediate the *sensation* of peripheral movement. Only *insofar as we feel them* can we carry out voluntary movements.[88] Let us look somewhat more closely at this relation between self-movement and sensation.

b. MUSCLES AND NERVES

Galen had distinguished between two kinds of movement: the involuntary *motus naturales* of the organs and the voluntary *motus animales* of the muscles, the former deriving from the *vital spirits* of the arteries and the latter from the *animal spirits* of the nerves; thus, heart and brain represent the two sources of movement. On this view, then, the muscles receive their power of movement through the hollow nerves from the brain: "Like rivers . . . the nerves bring their powers to the muscles from the brain as from a spring."[89]

In contrast, we have seen how Harvey, who reduces the various spirits to a special quality of the blood, correspondingly also sees all power of movement as proceeding from the blood system. In this, he is returning in a certain sense to Aristotle. However, while Aristotle reduces every *movement itself* to an alteration of the heart determined by temperature, Harvey stresses the contractibility and autonomy of the muscle. To the nerves, he assigns the function of triggering and modulation. Thus, muscle occupies an independent position between heart and brain and is the mere organ of execution of neither the one nor the other. Muscle receives a further reevaluation through the fact that, in line with his unifying tendencies, Harvey now also views the involuntary movements of the organs, which are thus no longer centrally controlled—above all the heart itself!—as muscular (*nullum motum sine musculo, intestinorum, cordis, pupillae*; p. 108). Chapter 16, *Nervi Usus in Musculis,* is to be seen against the background of these departures from Galen as well as from Aristotle.

As we are told there, the *capacity* for sensation and movement proceed from the heart along with the spirits. The *actualization* of movement then occurs through what is "sensitive in itself," not primarily through the brain; that is to say: *muscles exhibit a primary sensibility that permits the triggering of their movement even without the help of the nerves.*[90] Harvey demonstrates this through some examples: many muscles have no nerves at all, others move themselves although they are only simply innervated, yet differentiated in their parts, and so on. But, above all, in the vegetative sphere, muscular movement is grounded in the organ itself and not in the nerves. In another passage, Harvey also speaks of the sense-dependent, involuntary motion of vegetative organs, which is not induced by the nerves.[91] True, Galen had also attributed to these organs a kind of "irritability", but only under pathological conditions: disruptive materials call up the *facultas expellens* from hollow organs like the stomach or the gall bladder and stimulate them to expulsive contraction.[92] Here, too, Harvey turns pathological into physiological processes: *sensus* understood as irritability is, as will become still clearer, the presupposition for movement as such.

Muscular movement independent of the nerves is expedient and physiological in those cases where constant actions need not be capable of "voluntary modulation"—thus, for example, with the heart or the intestines.[93] A *pathological* self-movement occurs, on the other hand, when an *otherwise existing* nervous control fails: with decapitated animals or with human beings in delirium, where the muscles, now overmobile, fall into uncoordinated convulsions (ibid., p. 102). In this very way, however, the fundamental *autonomy of muscle* is revealed: "Muscles are as it were separate living things"; their elementary self-motion is a quivering pulsation—like that of the blood in embryogenesis or that of the *heart*, which Aristotle too, after all, has described as an autonomous living thing (*musculi esse tanquam animal seperatum* [sic] *qui cum in actu sunt quasi palpitant ut cor Aristotelis quasi animal seperatum* [p. 110]).

But in order to transform this self-motion into an ordered process, there is need of *sensation*—either of the *sensus* that is peculiar to the vegetative organs themselves (it will become evident later that order here depends on a stimulus-response relation) or of the *sensus* that communicates with the brain through the mediation of the nerves and so makes possible purposive voluntary movements.[94] Thus, nerves are the instrument of the common sense (*organum sensus communis*; ibid.) in contrast to the autonomous sensibility of the organs.

Harvey explains in more detail how order arises in voluntary motion. The nerves are neither themselves the movers (*moventes motores*; p. 108), nor do they pass on the "command" of the brain (*imperium differe*; ibid.). Harvey disagrees here with contemporary ideas. Instead, the nerve modulates the muscular self-motion as through the "*intervention of a judge*" so that rhythm and harmony arise (*nervus ut interventum iudicis, opera per rithmum et harmoniam fiant*; ibid.). In fact, this self-motion is itself *something perceptible*. Insofar as the nerve transmits its perception, it permits the brain to make a "judgment," that is, the selection of purposive movements. Cutting of the nerves, on the contrary, leads to powerful, but inappropriate muscular movements.[95]

Thus, Harvey arrives at a positive answer to the question posed in chapter 14, whether the motor nerves are at the same time sensitive: *the nerve of a muscle is the perceptual organ for its movement*, or as Harvey had already remarked in that chapter: "The movement proceeding from the brain . . . [is] identical with the sensation arriving at the brain" (p. 100). The brain is like a "choirmaster" (*cerebrum tanquam mester del choro*; p. 110) who permits rhythm and harmony to arise, insofar as he is able to hear every voice and merely gives appropriate directions to the singers. The muscles wait for the choirmaster's signal; they are ready for contraction (*musculum esse aptum ad contractionem*; p. 114). Their initial tension is manifested by their autonomous contraction when they are separated or lose their support through fractures (ibid.). Thus orderly movement arises through the interplay of contractility, triggering, perception, and modulation.

The last chapter of Harvey's work is again devoted to this theme. It compares the arrangement of the muscles with a well-ordered state in which not every single action requires the presence of the monarch (p. 146).[96] Instead of this, the muscles are organized as independent units of movement, which are set into motion, not through an equivalent cause in the mechanical sense, but already through slight "signs and watchwords" (ibid.). The harmonious reciprocal intervention of muscular actions, the equilibrium in the alternation of rest and movement, offer a "wonderful spectacle," a "silent music" (*admirabilis speculatio Harmonia et Rithmo musculorum de proportione moventium manentium. Tacita musica*; p. 146). The essential characteristic of this music is the rhythm that is formed, not by a metronome, but by a *relation of reciprocity* (*vicissitudo*; *ibid.*)

The last reflections in the text are also concerned with this relationship (*vicissitudo motus musculorum ad rithmum*; p.152). Is the reciprocal relation to be thought of as being like a state of equilibrium, or as like the deflection of the compass needle in the proximity of a magnet? Or is it as Aristotle conceived cardiac motion, namely, a consequence of the boiling up of the blood? (That would be only *one* principle of movement.) Or is it a reciprocal moving and being moved as with the muscles? In that case, relaxation would be part of the whole movement just as the diastole of the heart is complementary to systole.[97] Here, Harvey once again relates the problem of orderly movement expressly to the heart without, however, coming to a conclusive result. Voluntary muscular movement may indeed contribute much to the understanding of cardiac action, but this does not yet settle the question of the character of the *sensus* of vegetative organs and of a triggering and modulation of movement independent of the "judgment" of the brain. It was only later that Harvey pursued his reflections on *vicissitudo* until he obtained an answer to that question.

c. *SENSUS* AND *MOTUS* IN HARVEY'S LATER WRITINGS

First let us consider a section in chapter 57 of the *De Generatione* that takes up the theme of *sensus* and *motus* from the time of their origin (DG 256ff., W. 430ff.). From his observations and experiments with the chicken embryo, Harvey infers not only an elementary self-motion, but also a *capacity for sensation* in the blood (*in prima statim sanguinis guttala in ovo . . . sensum motumque clare elucescere*; 256, W. 430).[98] Just so does the embryo, as early as in its first, simplest stages, react to mechanical stimuli with wormlike contractions (ibid.).

Thus, Harvey says, if *sensus* and *motus* already appear before the formation of the brain, then the later involuntary or natural movements of the organs must also be conceived in this way: they are based on an *elementary sensibility of the tissues* that is independent of the brain.[99] So, the heart reacts to disturbing influences with disturbances of its rhythm, the stomach with cramps and vomiting. The flesh and the skin itself are able to distinguish from one another harmless and harmful (e.g. poisonous) influences and to meet the latter with contractions and rejection (257f., W. 431). But wherever we observe a reaction to irritation, we must also assume sensibility; indeed, this reaction is precisely the criterion by which we distinguish the living from the dead.[100]

Harvey continues: just as the five senses at first perceive independently, and then send what is perceived on to the brain, so there exists everywhere in the tissues a primary, natural sense or a kind of *touching*, which is not reported to the *sensorium commune* and so remains unconscious.[101] This elementary sensation, Harvey tells us, is also exhibited by the lower animals that lack a brain, such as worms, caterpillars, or zoophytes (*ibid.*). Thus again, we meet here the phenomenon of *hierarchical organization*: in the course of phylogenesis and ontogenesis, the original vital sensibility and mobility are taken up, modified, and ordered—in this case, through the brain.

This holds in the last analysis for the muscles as well: through irritation, they can be put into spasms and convulsions—a phenomenon which occurs especially after the lapse of central nervous control (Harvey summarizes here the results of the *De Motu Locali*). They then revert, as it were, into the early embryonic state (see above.) Only through the *sensorium commune* is the elementary sensibility of the muscles assimilated (and later also transferred to consciousness), and thus has its contractility directed into orderly paths.[102]

Harvey once more demonstrates this relationship with the example of the lung (260, W. 433f.). Its movement is natural and involuntary, and it also reacts to irritations (namely, with coughing or quickened breathing). But, Harvey says, from this self-motion and sensibility, under the guidance of the brain, the human voice, even human song, takes shape. Just as the transformation of the primal motion of the blood through cardiac action was a mechanical "trick" of nature, so art itself begins with the cerebral transformation and modulation of the *sensus* and *motus* of the tissues.

In the *De Motu Locali*, Harvey has already determined that stimulation of muscular sensibility and thus the triggering of a contraction can also take place through expansion (*[caro] facit contractionem dum distenditur*, DML 88). Here we have the solution to the riddle of *cardiac action*, which Harvey has already described in the *De Motu Cordis* as a kind of "answer".

Expansion, as we are told in chapter 51 of the *De Generatione*, is the *earlier* of the two motions of the heart, but this occurs through the *blood* from which cardiac action thus arises.[103] Harvey means by this not only the streaming of the blood into the auricles, but its regular *expansion* as the effect of its inherent *calor* or *spiritus*:

> Diastole . . . occurs through the blood, which is swelling in virtue of its inner *spiritus*, so that Aristotle's opinion about the heart beat (namely, that it occurs through an ebullition) is true up to a point.[104]

This swelling is an irritation appropriate for making the auricle contract and thus releasing the automatism of cardiac motion. For the heart is formed as an auxiliary organ for perception, transposition, and strengthening of the blood's own movement (*instrumentum ei usui destinatum; ibid.*).

> Thus it is certain that that little vesicle [in the egg; auth.] and later the auricle (from which the pulse begins) is incited by the expanding blood to the movement of contraction.[105]

The same process occurs in *expanded* form in the ventricles: expansion through the blood streaming in stimulates them to contract.[106]

In the second disquisition against Riolan, in which Harvey repeats his representation of cardiac action without essential change, diastole is consistently described as an impulsion, even as a "violent motion."[107] The relation between the motion of heart and blood as we have learned to know it in the *De Motu Cordis* is here reversed: *the blood now overcomes the natural tendency to movement, namely, the contractility of the heart*. In this way, the contraction or activity of the heart becomes the passivity of the blood while the activity of the blood becomes the relaxation and passivity of the heart. Just as flexion and extension are carried out by antagonistic muscles, so must systole and diastole be due to two different agents (ER *ibid.*). The circulation of the blood and the life of the developed organism are grounded in this mutual influence and in the reciprocal overcoming of two polar tendencies to movement.[108]

Insofar as it does not result from "selection" through cerebral guidance, orderly, rhythmic motion arises from an interrelation that presupposes the ability to move as well as to perceive. The basis of the *vicissitudo* in cardiac activity is the blood; in order to make its role an active one even in the developed organism, Harvey has recourse to Aristotelian *ebullitio*.[109] However, it is not the heat of the heart that allows the blood to swell—the heart itself receives its heat only from the blood of the coronary arteries as the necessary condition for its contraction (DG 203, W. 377); rather, what allows the swelling is the newly kindled *internal heat* of the blood itself gathering before and in the heart. In line with ancient precedents, Harvey likens this process to *fermentation* as the effect of inner heat.[110]

Thus, the "rising and falling of the blood" or the rhythm of cardiac action does not depend on an external principle or on a kind of metronome. The case is rather that the interrelation of perception and movement is variable in itself; it is governed by "an inner, regulating principle," in the last analysis, the soul itself.[111] Here, then, is the place at which the closed system of the circulation can be open to the influences of varying blood flow, processes of disease, and psychological conditions—influences that, as Harvey describes them elsewhere, modify the pulse and rhythm of the heart and the quantity of blood per beat.[112] Granted, Harvey gives no fuller elaboration of the exact way in which the circulation is regulated; the treatment of this theme promised in the *De Motu Cordis* (60, W. 50) was never realized. But that the motion of heart and blood form a harmoniously adjusted, subtly reacting whole is apparent from many passages. Moreover, Harvey's introduction of the soul as regulating principle precisely at this juncture—the only time it is mentioned in connection with the circulation—shows that the *vicissitudo* of blood and auricular musculature represents the central point of the circulatory process.[113]

This higher principle can be effective because, and only because, *two contrary vital tendencies to move here condition one another and hold the scale*—in contradistinction, therefore, to a uniform "motor" of movement. As early as the *De Motu Cordis,* Harvey had spoken of the "harmony and rhythm" of cardiac action. His reflections at the end of the *De Motu Locali Animalium* also tended in the same direction: he asks whether the *vicisssitudo* resembles a state of equilibrium or is like the compass needle; does it consist in an alternating enlargement and diminution of the parts? It now becomes evident that as a scale in equilibrium perceptibly registers the slightest changes so does precisely the harmonious balance of the two poles of cardiac action—in its alternation of rest and motion, expansion and contraction, relaxation and hardening—make possible a living rhythm, which as a whole is again sensitive to changes in the organism, indeed which even "perceives" psychological phenomena and mirrors them in its movement.

If we now return once more to the *De Motu Locali*, we may perhaps understand still better Harvey's admiration for harmony and rhythm, the "silent music" of the organism:

> With the help of the muscles nature carries out her works in living things through rhythm and harmony. . . . By divine agency, it is clear, delightful and charming motions are produced in the heav-

ens for which we have no more sensibility than dogs do for music.[114]

d. SUMMARY: POLARITY AND MOVEMENT

Chapter II described the relation between heat and motion as a basis for the occurrence of the circulation. Now, as a third moment, chiefly present only by implication in the *De Motu Cordis*, we have sensibility, an elementary *power of sensation* on the part of the tissues, which allows them to respond to stimuli with an actualization of their *elementary tendency for movement*.

> ... sensation is there for the sake of movement, and movement is directed by sensation, both as to when it happens and also where and how and by what means (DML 104; cf. note C-88).

Thus now, the bodily heat emanating from the blood and heart is a presupposition not only for mobility, but also for the sensibility of the parts. Hence, the relation *calor > motus > calor* is expanded to the cycle *calor > sensus > motus > calor*.[115]

Living movement, though a *self*-motion (DML 18), still does not come about uncaused. Rather, it is "evoked" by its counterpart, i.e. by an antagonistic stimulating movement. Elementary sensibility and motility alone, however, lead only to uncoordinated reactions, to convulsions or arhythmic movements. Yet order is not forced on them from outside (for example, by an impulse giver inside the body), but is formed through the fact that what is sensitive *faces its counterpart in a relation of reciprocity*, which—as in the case of the pulse or of respiration—maintains itself and becomes stabilized as a *rhythm*. The configuration that has thus come about is itself sensitive and capable of modulation as a whole in contrast to a mechanical stroke. This makes possible a *hierarchical arrangement of elementary relations of movement*, which allows the circulatory process to become an expression of general bodily or of psychological events, and respiration even the expression of human art.

This holds in a similar way for muscular movement in general. Here too, there exists a synergism of protagonist and antagonist, a harmonious interaction of rest and movement. In this way, units of movement are formed, which can be triggered as wholes, namely, under the representation of the *goal* that they serve: "*Natura enim cogi-*

tat de operibus, non de motu musculorum, nature thinks of the work, not of the particular muscles (DML 122).[116] Again, the order is not simply imposed from above; the brain prescribes only the particular items. But above all, it is an *organ of perception* for the independent vital configuration of movement, that is, the *sensus communis*, which receives and integrates the peripheral *sensus*.

The heart, too, is a muscle; it has lost the special character of a *pulsative faculty*, and its motion is the reaction to stimuli—thus *effect, not cause*. Nevertheless, the heart remains independent of the brain, as in itself an *organ of perception* for the motion of the blood as well as for psychological influences. As an "inner animal" in its own right, it possesses *sensus* and *motus*.[117] These traits develop in it through its origin from the *punctum saliens*. Even the first drop of blood shows reactions; it "lives, moves itself and senses like an animal."[118]

But the *punctum saliens* pulsates without a *vicissitudo*! That is to say, *at the beginning of development the division of polarity is not yet present*, and this elementary pulsation recurs at the end as well, when, after the heart has stopped beating, the blood in the right auricle still exhibits movement (*inundatio et pulsatio*). In such places, Harvey always speaks of *calor* and *spiritus* as a presupposition of this movement (DMC 33f., W. 28; DG 68, W. 238, 200, W. 374, 202, W. 377); "the blood itself or spirit" carries this pulsation in itself (DMC 34, W. 29); "through the spirit pulses the life blood" (DML 94).

The two directions of movement of the Aristotelian *pneuma*, expansion and contraction, are thus united in the pulsing blood. With the differentiation of the organism, this polarity, too, develops into contrary moments of movement. Active expansion or dilation remains with the *blood*, which swells "as it were, in consequence of containing an inherent spirit"; this corresponds to the *immediate* effect of the innate heat.[119] But contractility is transferred to the heart and to the musculature as a whole; Harvey speaks in this connection of the "motive spirit."[120] Heart and muscle also need the innate heat, if only *mediately*, for the maintenance of their contractility—but not to effect the contraction itself by means of heat.

Thus in Harvey, the principle of vital tendencies to movement is derived from the spirits as a single moving force, which, however, assumes particular directions in the tissues. Contraction and dilation remain an *alloiosis* or alteration, and thus qualitative changes made possible through heat. Nevertheless, they are no longer the effects of a single *pneuma* that transforms hot and cold into contrary motions, but are instead specific actions of the blood and muscle tissues themselves.

Yet in this way, the elementary polarity of the blood is transformed into a reciprocal relation. It is not the blood alone that is responsible for cardiac action as it was in Aristotle; nor is it only the heart itself as Galen had taught. Harvey knows well the phenomenon of the empty heart that goes on beating, which he describes in the *De Motu Cordis* in the case of an excised fish heart (DMC 33, W. 28). But this corresponds only to the elementary sensibility and contractility of the heart, which is also displayed in isolated muscle fragments of these fishes (*ibid.*); and it no more contradicts the *vicissitudo* in the *living* organism than it does the blood that continues to pulsate after death.

Renouncing the unified, all-explanatory power of movement of the *spiritus* and the *facultas pulsativa* or *motrix* of the soul in the heart, Harvey develops a new conception of movement that is no longer purely ancient nor yet mechanistic, but one that is based on polarity, antagonism, and reciprocity. Last but not least, we see here the influence of renaissance ideas of nature: harmony and rhythm in the organism correspond, as Harvey says, not only to music, but also to the operation of *sympathy and antipathy* in the world; from these, activity and passivity in living movement are in turn derived.[121] Movement becomes a union of antagonism and cooperation.

IV. DE GENERATIONE: THE PRINCIPLES OF THE CIRCULATION

<u>Summary</u>. *Living movement as reciprocal relation—that is for Harvey not yet the ultimate level of scientific knowledge. There must be an initial stage in which this polarity still displays an undifferentiated unity: the Aristotelian "arché," the first organ to develop and to move itself. But in it a first moving principle must be operative—the "primum efficiens" of the conception, development, and movement of the organism.*

The following chapter investigates this problematic in Harvey's magnum opus, the De Generatione. *In the first place, it considers the account of early embryogenesis presented there. (a) Harvey sees in the punctum saliens not, as was traditional, the heart, but rather the blood as self-moving fluid and thus as the primary and basic form (arché) of the heart and circulation. (b) Accordingly, organic development does not take place on the basis of preformed material structures as contemporary preformationist embryology taught. In its place, Harvey sets "epigenesis," the increasing consolidation and hierarchical organization of the originally formless fluid. In Harvey's view, partial structures, even elements or atoms, can arise only secondarily as "precipitations" of living organic substance.*

A further section, (c), considers the double aspect of the blood. On the surface, it is "nourishment," material substance, but beneath that it is of a high-

er, divine nature. Transformed into spirit or innate heat it becomes the very principle of movement of life itself. (d) At a still deeper level, Harvey inquires into the origin of this second, superior aspect. At the same, time this is the question of the ultimate, unitary principle of the conception, development, and circulatory movement (that is, the autonomy) of the organism—the question, in other words, that brings the De Generatione *and the* De Motu Cordis *together as a unified whole. Harvey identifies this principle with the divine cosmic nature itself, which reaches down into the vegetative soul of the living being; thus this "world soul" ranks higher than even the individual rational soul of a human being.*

Finally, a concluding section, (e), contrasts "cycle" and "polarity" to one another as concepts characteristic of Harvey's thought. Polarity—like that of male and female, solid and fluid, heart and blood, center and periphery—lies at the basis of the various cycles as condition and stimulus. Between these poles there reigns not only a productive tension, but also a "gradient" so that one pole is placed before and above the other (for example, the female before the male, the blood before the heart, the center before the periphery). In the last analysis, polarity shows itself to be derived in each case from an original unity, but this unity is re-established through the cyclical process, the cycles of generation and circulation in nature and in the organism.

As early as the *De Motu Cordis,* Harvey referred to his embryological investigations for the clarification of the question of the principle of the motion of heart and blood.[122] The significance of the developmental approach for Harvey has already become evident a number of times in the phenomenon of hierarchical organization. The differentiated organs and processes of the developed organism often conceal their true basis. What is closer and better known to us often represents a variously compounded and transformed multiplicity, which has to be analyzed in order to reveal its principles. But these principles are best exhibited in their pure form at the beginning of development.

Now, according to Aristotelian doctrine, precisely this goal of development, the whole organism, must already be present as a *principle* from the start. Thus, we are seeking the *arché* of the organism, that is, not only the principle from which the further movement of ontogenesis proceeds, but also the principle that already contains within itself the organism *in nuce.*

> Therefore some part must exist first, in which the source of motion lies, for this is from the beginning a part of the end and the most important thing. . . . Thus if a part is so fashioned for animals, including the beginning and the end of the

whole being, then it must develop first and arise at the start as the source of motion.[123]

In what follows, Aristotle identifies this principle with the heart, which for him, too, is like a living thing in its own right.[124] Thus here the conception of two opposing orders of development, the ontogenetic and the ontological undergoes an important modification. Although the developed organism, as the end and thus the higher in the level of being and *ontologically* "earlier," is realized only at the close of development as pre-form or prototype, it is nevertheless anticipated *ontogenetically* from the start and thus physically as well. The entelechy makes use of a bridgehead, so to speak, which represents it at the physical level so that the organism can subsequently be developed from it. We shall see that Harvey adopts this conception, reinterprets it, and even raadicalizes it.

In order to follow Harvey's view of ontogenesis and its principles, we shall turn first to its initial stages in order then to investigate its law-like regularities, its *arché*, and finally, the ultimate causes of the organism.

a. THE EARLY STAGES OF ONTOGENESIS

Following the embryological tradition ever since Aristotle, Harvey describes the development of the fertilized hen's egg; what is true of it, according to his observations (especially in deer embryos, DG 308ff., W. 482ff.), holds analogously for the development of animal and human embryos in general.[125]

As its starting point, Harvey recognizes the *cicatricula* or blastoderm in which two or three concentric circles are to be seen on the second day of incubation (DG 62f., W. 232). On the third day, there appears in its center a fully transparent fluid, "clearer than the water of the eye"; it has been produced from liquefied albumen, and is called by Harvey *colliquamentum*.[126] *Such a globular, membrane-enclosed fluid is for him the beginning of all living things.* He calls it *ovum* or *primordium*, and the fluid itself he calls *humor primigenius*.[127]

On the fourth day, there appears in the center of this fluid the speck of blood that we have already mentioned several times—a speck that now begins to beat rhythmically. In this speck, Aristotle already saw the heart; according to Harvey, on the contrary, the heart forms only later around the *punctum saliens*; and even before the pulse of the latter, the blood itself exists.[128] Since he comes here to the limit of his powers of observation—he had only a magnifying glass at his disposal (72, W. 241)—Harvey brings reflection to his aid: the application of heat can often stimulate the already motionless drop of blood to move so long, that is, as it itself still contains *calor innatus*— or, in other words, the vital heat of the blood is the necessary condition for every movement, even for its own pulsation.[129]

However, as Harvey tells us, pulsation signifies *the transition from quasi-plantlike to animal life*. If up to now only transformational and developmental processes resulting from a vegetative soul were to be seen with this autonomous rhythm, the inner production of heat and thus the self-maintenance of the blood begin, in other words, *the autonomy of animal life*.[130] Perfected in the developed circulation, it nevertheless manifests itself first of all in the elementary movement and sensation of the *punctum saliens*. Its rhythm forms itself even without stimuli, spontaneously, as self-perception of the blood, so to speak; it already exhibits its sensitivity to irritation through change of amplitude or frequency (DG 69, W. 239). "It is not to be questioned, therefore, that this punctum lives, moves, and feels like an animal (*ibid.*)."

However, not only a power of movement and of perception, but also the *vis plastica* (60, W. 229; 66, W. 235), the psychological principle that "fashions the chick from the egg," begins to operate with the appearance of the *punctum saliens* (*animam pulli ingredi, quae ex ovo pullum format, eumque postea informat*; 69, W. 239). For as a consequence of the primary motion of the blood, fine vessels make their appearance in the colliquament (69, W. 238) in the course of the fourth day; and thus, the structures of the organism are gradually formed as *parenchyme* ("something poured on the side"), as deposits or, as Harvey also says, *coagulations* of the blood streaming out of the *punctum saliens*.[131]

The first drop of blood had arisen from the colliquament under the influence of the heat of incubation. Now, the heat of the blood in its turn serves for the further "concoction" and transformation (*concoctio, transformatio*, 134, W. 305, 153f., W. 326) first of the colliquament and later of the egg white and the yolk in order to make more and more new blood available as nourishment for the growing tissues (73, W. 242). With its heat as the "common instrument of all vegetative processes," the blood is "both a part . . . , and a kind and efficient [cause] or instrument" of the emerging organism; blood and *punctum saliens* "attach round themselves the rest of the body, and all the other members. . . ."[132]

With the appearance of the vessels, the formation of the heart also begins. At the end of the fourth day, the *punctum saliens* develops into a spontaneously contracting vesicle that moves the blood (69, W. 238). This, Harvey declares, is the necessary sequence: the blood must exist before its receptacle, the content before the container that serves it (*quod continet, contenti gratia fabrefactum esse; ideoque posterius efformari*; 85, W. 255, also 200, W. 374). But the first blood vessels as well precede the formation of the vesicle: only when the blood needs a central impulse for its longer paths does the antagonism between expansion of blood and cardiac contraction come into play.[133]

Next, the primary vesicle differentiates into two vesicles with opposing pulses (71, W. 241) of which one first forms the auricles and the other, later, the ventricles. Thus, the heart arises through the *polar development and gradual parenchymal transformation of the* punctum saliens.[134] At first, it continues to be external to the developing body, its "future dwelling place," until finally it moves in and closes the thorax around itself.[135]

Even at its first appearance, the blood already possesses movement and sensation like an animal. It is the part in whose initial existence the organism as a whole is already anticipated, the part that allows the organism to emerge from it and which brings it to actuality (*veluti totum ex parte*; 154, W. 226)—in a word, its *arché*. There are no solid parts at the beginning of development, neither the trio of liver, heart, and brain as in Galenic embryology, nor a central organ as in the Aristotelian version. *Fluid itself, by its very movement, shapes the structures according to whose outlines the organs form as deposits and consolidations.*[136]

Even in mature organisms, the organs still continue to depend on the ubiquitous presence of the blood, of the *particula genitalis prima;* its constant maintenance and regeneration is itself the goal of bodily processes. Thus, ontogenesis confirms and explains the identification of the efficient and final causes of the circulation in the De Motu Cordis: "The part, in a word, in which inhere both the principle whence motion is derived, and the end of that motion, is obviously father and sovereign."[137]

b. THE REGULARITIES OF DEVELOPMENT

As we have seen, embryogenesis takes place in a twofold process of transformation: from the colliquament (or egg white and yolk) to the blood, from the blood to the parts and organs. Let us try to understand more closely the inner principle of this development.

In this context, Harvey frequently uses the concept of *law*: organs are differentiated out of unorganized material "under the law of generation" (*procedente generationis lege*; 343, W. 517); it is a "law of nature" that parts take shape only when they have a function, or that the content comes into being before the container (*ex lege natura est*, 236, W. 410; and 241, W. 415); nature and the formative principle "operate in conformity with determinate law" (*secundum leges operante*; 195, W. 369). As we can see, it is not a question of "laws of nature" in the modern sense of external boundary conditions of the physical world, but, conversely, of inherent principles of development to which matter is subordinate. Nature prescribes its norm to the instruments of generation (*natura . . . [iis] operandi normam praescribit*; 191, W. 365).

To this corresponds Harvey's fundamental rejection of contemporary atomistic and preformistic embryology, which tried to explain development as the assemblage of different building blocks, parts, or rudiments, that is, by reference to the materials it started from. ". . . [T]hey who philosophize in this way, assign a material cause [for generation], and deduce the causes of natural things either from the elements concurring spontaneously or accidentally, or from atoms variously arranged."[138] But the development of the living is more than "a separation, or aggregation, or disposition of things"; it cannot be conceived without a higher efficient cause gifted with "foresight."[139]

In fact, Harvey declares, we never observe the putting together of preexisting heterogeneous components (342f, W. 516f), but, on the contrary, their formation out of a common, homogeneous basic material whose potentialities are realized, not through local motion, but through qualitative alterations (*alteratio, transmutatio*; 235, W. 409; 243, W. 417). Inner heat is the instrument by which the *vis plastica* or the *internum principium motivum* subjects this general material substrate to change of form and reshapes it into colliquament, blood, and finally, organs (153f., W. 325f.)—just as the same rain can assume quite diverse properties and determinations of form in the plants that it nourishes (bitter, sharp, sweet etc.), and so actualizes its potentiality according to the particular substance formed in each case.[140]

For this development of form through successive transformation of the underlying material, Harvey introduces the concept of *epigenesis*; this term denotes the simultaneous process of taking shape and of growth in contrast to a mere enlargement of preformed parts.[141] ". . . [A]s when the potter educes a form out of clay by the addition of parts, or increasing its mass, and giving it a figure, at the same time that he provides the material, which he prepares, adapts, and applies to his work, . . . so exactly is it with regard to the generation of animals."[142]

Epigenesis means the superimposition of increasingly more

solid parts on elementary structures. What arises first is the *arché* or core to which the particular organs and limbs are added in prescribed order; "the edifice of the body . . . is raised on the *punctum saliens as a foundation*."[143] Further, it is a "law" of this development that it produces the parts only when they also have a function (236, W. 410). Hence, the model for epigenesis is the heart, which develops from the blood that already exists and from its movement, as soon as the size of the organism demands it; for the blood "uses the heart as instrument" for its expanded movement. The blood is earlier "both in the order of generation and of nature and essence," thus in the order of ontogeny as well as of ontology. The heart serves the blood rather than *vice versa*. In the springing point of blood, Harvey sees the kernel that anticipates *in nuce* the heart and the organism as a whole.[144]

> There is a greater and more divine mystery in the generation of animals than the simple collecting together, alteration, and composition of a whole out of parts would seem to imply; inasmuch as here, the whole has a separate constitution and existence before its parts, the mixture before the elements.[145]

That both expressions, the "whole before the parts" and the "mixture before the elements" are intended here in a more radical sense, that is, both logically and materially, is clear from chapter 72, "De Humido Primigenio."

Humidum primigenium is what Harvey calls the "crystal-clear colliquament," the simple, pure elementary fluid from which the blood is formed and which contains all the parts potentially in itself. For living things, it has the characteristics of the *prote hyle:* itself unformed, pure potency, it can receive all forms. As the eye can see only because it is itself wholly transparent, as the mind as "potential intellect" is able to receive the forms of all things without the matter, being *forma formarum*, so something must precede and lie at the basis of epigenesis, something that is itself without any structure or composition, the *stuff of life in pure potentiality*.[146]

Hence, Harvey refuses to take purely external aggregation and ordering of elements or atoms as the beginning of development. It is, he holds, a widespread error to view the homoiomeres—since Aristotle, "similar parts" had been the tissues (flesh, sinews, bones, etc.) of which organisms were composed—as agglomerates of differentiated particles. In fact, such an assemblage is never to be seen in the development of organisms (342, W. 516). On the contrary, ontogenesis proceeds from a fully homogeneous, gelatinous fluid. Its qualitative transformation under the "law of generation," not than the local motion of particles, brings about articulation and differentiation and allows the "organic" to arise "from the inorganic."[147]

According to Aristotle, the primary bodies or elements must be "for the sake of the homoiomeres, . . . *since the latter follow the former in their origin*" (DPA II,1, 646b6ff.). For Harvey, in contrast, the vital precedes the particular elements not only in rank, but also *temporally and physically*. The life stuff and the tissues differentiated from it contain no mixed elements prior to themselves, but rather exist before their elements (these, according to Empedocles and Aristotle, being fire, air, earth, and water; according to chemists, salt, sulfur, and mercury; according to Democritus, certain atoms) as being naturally more perfect than these.[148]

Only death and the dissolution of living substance allow elementary materials to arise as do burning, artificial dissection, and alchemical distillation (344, W. 518). But all these are *end states*, residues, or sedimentations that have escaped from the material unity of the living—they could never become, in the reverse direction, principles of the living.[149]

According to Harvey, the stuff of life and the tissues *as such* can never be decomposed into elements in actuality, but only in thought (*dissolvuntur . . . in ista, ratione potius, quam re ipsa et actu*; 343, W. 517). Aristotle, too, seems to speak rather of the potential than the real existence of the elements. Viewed from the perspective of life, all particles, atoms, or elements are merely *entia rationis*—only death produces them. The living, in contrast, arises only from the living; the development of organisms follows its own laws and cannot be understood in terms of what is dead.

Harvey is well aware that with his point of view he is placing himself in opposition to the dominant tendency of contemporary thought. The lines from a later letter express ironic resignation:

> . . . nor do I doubt but that many things still lie hidden in Democritus's well that are destined to be drawn up into the light by the indefatigable diligence of coming ages.[150]

c. THE BLOOD AS PRINCIPLE

> The life, therefore, resides in the blood, (as we are also informed in our sacred writings), because in it life and the soul first show themselves, and last become extinct.[151]

The entelechy of the organism is present primarily not in an organ or in a mixture of parts, but in the elementary stuff of life itself. The blood is the *arché*, the *primum vivens ultimum moriens*, the source of autonomous organic heat. It contains in itself the polarity of expansion and contraction as well as that of perception and motion (*sensum . . . et motum sanguini inesse*; 207, W. 381). On this depend not only, as we have seen, the pulse and the later action of the heart, but in general all vegetative and sensitive processes; the blood as the source and origin of all the members is also present and life-giving in every part of the body, however minute (205, W. 377; 203, W. 379). *It lives of itself* in contrast to its containers—the heart and blood vessels; indeed, the whole body remains, as it were, an envelope, an "exudation" of the blood, constantly dependent on it (*sanguinem per se vivere . . . totum corpus ab ipso dependeat*, 205f., W. 380).

The blood itself is therefore the basic element of the body, the starting point of development, and its goal. As unity of material, efficient, and final cause, it becomes the vital principle as such, the *vinculum animae*, indeed the *first seat of the soul itself*.[152] Thus, the blood is not only the chief instrument of the vegetative faculty (201, W. 376), not only the bearer of vital power and health, or of their opposite, sickness and death (206, W. 380); the flow of the blood is also the *immediate expression of moods and feelings*.

We have already seen in the *De Motu Cordis* how the blood gathers as a whole and contracts at the first stirrings of fear, terror, and the like; this centralizing movement is also mentioned in the *De Motu Locali* (*concentratio sanguinis in passionibus*, DML 102). In addition, the *De Generatione* describes the contrary motion of the blood to the periphery, namely in joy or anger (DG 203, W. 377). In the second discourse against Riolan, Harvey again speaks extensively about the psychological influences on blood flow: fear stops the flow of blood, shame drives it into the cheeks, illness make them grow pale, and so on.[153] Thus, every movement of the soul is expressed in the motion of the blood, which then, through its reciprocal relation with the heart, influences the pulse and cardiac rhythm as well.[154]

Even in the form of the first drop, the blood is an "independent organism." It also preserves this initial unity in the circulation, in which, as a closed, living whole, it moves, reacts, and represents emotional states. But the *homogeneity* of the primary stuff of life is also maintained. Harvey criticizes the Galenic and Aristotelian doctrines of distinctive component parts of the blood, which were based, one-sidedly, on the observation of "extravasated" and coagulated blood (212, W. 387). On the contrary, as a living, ensouled body part, the blood has everywhere the homogeneous constitution of a *pars similaris*, a uniform stuff (213f., W. 387f.). Only outside the vessels, when heat, the principle of life, has left it, does it break up into the component parts of *serum, coagulum, ichor, sanies*, etc.; however, these parts have no special existence in the living blood, but appear only when it is spoiled and dead.[155]

Thus, the blood has a double aspect (*sanguinis igitur, prout est corporis pars vivens, ambiguae naturae est, et duobus modis considerandus venit*, 216, W. 319). As stuff subject to motion, it is nourishment for the body; as bearer of heat and spirit, it is the body's vital principle. This double character of the blood is justified at greater length in chapter 71 and in the polemic against Riolan. Blood without heat and spirit becomes immobile, inanimate; it is no longer blood, but *cruor*, dead coagulated substance (DG 328, W. 502). Just as a dead hand or one made of stone can no longer be called a hand in the proper sense, so blood without its vital principle is no longer blood; spirit and heat belong to its very concept, they are its reality, *actus*, and essence.[156]

Fernel and the other physicians assume special spirits for each of the essential bodily functions, not only the natural, vital, and animal spirits, but also visive spirits, auditory, concoctive, aerial, ethereal, and so on. But the blood alone fulfills all these functions, and we are not to assume more explanatory principles than necessary (*entia multiplicare sit supervacaneum*, DG 328, W. 502; mistranslated by Willis). The spirits are supposed to be an ethereal stuff of celestial nature, nourished by the breath. But aside from the fact that this stuff is not perceptible by the senses, how are these divine properties to come from the ordinary element, air (329f., W. 503f.)? And if air cools, how can it undertake the formation of heat and spirit (ER 138, W. 118)? The blood itself has all the properties attributed to the spirits: subtlety, the highest mobility, and omnipresence in the body (DG 330f., W. 504).

According to Harvey, the deepest reason for the postulation of special spirits lies in the failure to understand that a visible and tangible material like the blood could have "more than material" properties and thus produce effects that transcend the capacities of its elements (*ea potissimum ratione nixi, quod sanguis [utpote ex elementis compositus] supra vires elementorum . . . agere nequeat*, 329, W. 503). If that were not so, one would indeed have need of a *deus ex machina* and would have to snatch the spirits down from heaven in order to explain the miracle of the conception and development of the living. But "what we are seeking in the stars, exists at home":

> The blood . . . acts with powers superior to the powers of the elements, . . . in the forms of primordial [part] and innate heat . . . , and [in] its producing all the other parts of the body in succession; proceeding at all times with such foresight and understanding, and with definite ends in view, as if it employed reasoning in its acts.[157]

As bearer of the principle of the soul with its ability to plan, the blood is spirit and hence *itself of divine nature* (333, W. 507). Indeed, all natural bodies fall under two aspects, a material aspect of elemental character and a higher one—insofar as they are governed by a superior principle. While a "primary agent" evokes no more than material consequences, an "instrumental agent" is able, in virtue of that higher principle, as it were, to exceed itself.[158]

This transformation makes of the blood more than a warm fluid. Circling in its vessels, equipped with innate heat, it becomes the shaping and animating principle of the body, a unique organ, indeed the immediate organ of the soul itself. Thus, it obtains the attributes that Harvey had also assigned to the heart in the De Motu Cordis: "sun of the microcosm," "hearth and house god of the body." In maintaining itself through its circular motion, it also maintains the parts with which it has surrounded itself—no differently from the way that the sun and moon in their circular motion warm and animate the lower heavens.[159]

The lower world is bound with the upper, from which all motion takes its origin (344, W. 508). According to Aristotle, the *pneuma* corresponds in its nature to the primal stuff of the celestial bodies, the eternally circling ether, the "quintessence". This characteristic is now transferred to the circulating blood: it is itself spirit and is analogous to the element of the stars as a divine instrument (*analogus elemento stellarum*, 336, W. 510; *coeli analogum, coeli instrumentum*, 334, W. 507). We should not stand in awe before names like "spirit" and "heat," says Harvey; instead, it is the concrete substance, *the visible and tangible, ordinary blood* that displays these divine attributes. Thus, the soul, too, is neither purely corporeal, nor purely incorporeal in nature, of terrestrial as well as of celestial origin (*anima . . . nec omnino corpus sit, nec plane sine corpore; partim foris adveniat, partim, domi nascatur;* 337, W. 511).

The miracle resides in the everyday; the higher world is always present in the lower. In the last analysis, it is the thought of *transsubstantiation* that Harvey is here transferring to the blood: "It therefore comes to the same thing, whether we say that the soul and the blood, or the blood with the soul, or the soul with the blood, performs all the acts in the animal organism."[160]

d. CIRCULATIO AND GENERATIO

Various kinds of connections can be established between Harvey's central themes, the process of circulation on the one hand and on the other, conception and development. Like Aristotle, Harvey sees a cyclical process in the succession of generations through which mortal individuals perpetuate themselves in the species; this circle begins and ends in the egg.[161] But in the life of the individual as well, nature returns in the end to the starting point: to the pulsing blood as the *primum vivens ultimum moriens*. In addition, the blood is the "first engendering part(icle)" (*particula genitalis prima*, DG 198, W. 373 (revision of Willis)), which, through its circular motion, both produces the members and animates and nourishes them—and nourishment is indeed "a kind of continuous (re-)generation" (*nutritio . . . generationis quaedam species est*, 336, W. 509). Finally, the circulation is also a qualitative cycle of regeneration of the blood itself, which is thus protected from blockage, transition to *cruor*, and dissolution into its component parts—just as reproduction prevents the dissolution of the species into mortal individuals.

Both generation and circulation imitate the movement of the celestial bodies, at once self-maintaining and life-giving (DMC 34f., W. 29; 57, W. 46; DG 115, W. 285). Therefore, what the blood is for the organism—namely, material, efficient, and final cause in one—the egg is as "pivot" of the sequence of generation. It is the goal that nature follows in individuals as well as the material and the efficient cause from which they arise (DG 101, W. 271f.). The egg is constructed like a "microcosm" (44, W. 213; 281, W. 455), and the concentric membranes of the *cicatricula* correspond to the spheres of the celestial bodies, whose movement the blood later imitates.

The analogy between circulation and generation could be carried further in the processes of cardiac action and of conception, which are both connected with pulsation and impetus, with the production of heat, swelling, and the transference of spirits.[162] But in all this it is not a question of *mere* analogies. For according to Harvey, there must be a *principle* that guides the eternal succession of generations (*aliquod principium esse istius revolutionis*; 115, W. 285). This principle makes the parents capable of reproduction and the egg fruitful; but it is also transferred from the egg to the blood, and thence into the heart and from the heart to the whole organism (174f., W. 347f.). It is one and the same principle, a *vis enthea*, that is present in these differing forms and that manifests itself as shaping, nourishing, or maintaining force while, "like a sort of Proteus" (115, W. 285f.), always remaining the same. Therefore, when we ask about the cause of all this—and science rests on the knowledge of the first causes (*scientia enim ex causis cognitis, praesertim primis, oritur*; 121, W. 291)—it is the same question whether we ask what makes the seed, the ovary, or the egg fruitful; what allows the parts to emerge from the *punctum saliens;* and finally, what shapes the living thing in its *Anlagen* and in its constitution: the question of the first efficient cause of generation as well as of circulation.[163]

We do not intend to follow here in further detail the way in which Harvey ascends "from the last to the first and supreme efficient cause" in debate with Aristotelian and contemporary doctrines of conception (*ab ultimo efficiente ad primum et supremum ascendere*; 188, W. 362 (revision of Willis)). He places the following condition on this "first efficient (cause)": it must supply efficacy, that is, fecundity, to all intermediate causes; it must contain the "plan" of the organism and thus the contribution of both parents to heredity; and it can evoke no response that stands higher than itself in the hierarchy of being.[164] For otherwise, it would be only an "instrumental efficient (cause)" like the sperm or the egg: independently effective, to be sure, but still subordinated to the law of the process as a whole and not empowered to produce the organism purely from itself.[165]

Would it then be, on the model of Aristotle ("man begets man"), the *parents* we are looking for?[166] They have indeed already realized in themselves the completed form and do not rank at a lower level than their descendants. But Harvey adds a further condition:

> . . . viz., that the prime efficient . . . makes use of artifice, and foresight, and wisdom, and goodness, and intelligence, which far surpass the powers of our rational soul to comprehend.[167]

But this foresight and organizing ability cannot be attributed to the parents, not even to the human parents, whose reason would not suffice, and who in any case participate in the act of conception not with their reason but with the *vegetative* power of the soul (194, W. 368).

Therefore, like seed and egg, father and mother, too, are only instruments of the highest foreseeing and planning cause: *nature, the world soul or the creator himself*—for all those terms have the same reference . . . *natura, vel anima mundi, vel Deus omnipotens [nam eodem haec redeunt]*; 193, W. 368). The search for the principles has arrived at God as the first efficient cause.

But God's first instrument is the *sun*. In its course through the ecliptic, it carries out an eternal and at the same time rhythmic movement of approach and withdrawal and thus becomes the cause of coming to be and passing away in the change of seasons on the earth (192f., W. 367). The inclination of the ecliptic changes the continuous, life-generating stream of heat into a rhythmic one: *polarity dominates even the first principle within nature, and it continues in all terrestrial circulations and movements*.[168]

Thus, one can also say with Aristotle: *the sun and man beget man* if what is meant is: the sun through man *as instrument*.[169] For only a planning, foreseeing principle that transcends man can produce him—the sun, that is, nature or God. Since this principle exceeds all the capacities of human reason, but uses the *vegetative soul* as the instrument of conception and development, the latter now undergoes a total revaluation over against the tradition whether Aristotelian or Christian: *it must be considered higher ranking and more divine than the rational part of the soul since it is, in the last analysis, a part of the divine nature itself.*[170]

The counterpart to the providence of nature and of the vegetative soul is the prudence of human reason (195, W. 369). The former operates out of inner lawfulness as it were blind and yet foreseeing; prudence, on the other hand, must take the detour of deliberation and the choice of means for the particular end it has in mind.[171] Nature is an autodidact, while man must learn; human art can imitate, only imperfectly and by stages, the formative power of the vegetative soul, the craftsmanship that animals possess through a *connatum ingenium*, or "innate intelligence." Only what has become "second nature" through discipline and habit acquires some similarity to the operation of God in nature (*cum habito perfecto in nobis existit, quasi altera anima . . . acquisita . . . effectus similes producit; ibid.*)

Thus it is not the conscious choice of goals, deliberation, and decision, but a kind of unconscious method, instinctive craftsmanship, even "blind foresight" that characterizes this operation. In this way, God is present everywhere in his creation; all natural objects are his work and at the same time his instruments (*eius Numinis et opera sunt, et instrumenta*, 196, W. 370). The blood as well, which forms all the organs, "proceeding at all times with . . . foresight and understanding, and with definite ends in view" (DG 333, W. 507), also has this power insofar as it is the "instrument of God" (334, W. 508). Thus, we can say "all things are full of Jove" or "all things are full of soul."[172]

For Harvey, the goal-directedness and systematicity of natural processes cannot find its ultimate explanation in individual beings. Precisely the events that characterize conception and the transference of the vital principle, which are so hard to penetrate, allow him to infer a single efficient cause that is present in reproduction as much as in the formation and maintenance of the organism through its inner movement. But in this way, his vital principle,[173] the vegetative soul, becomes a partial component of a world soul patterned on the Stoic model: *individuals in their vital aspect are organs of a universal nature.*[174] The Aristotelian-scholastic conception of the individual substance is dissolved since its limitation to the individual has been removed and substance itself has been universalized.

But if the world is an organism with God as its soul, could not then man with his individual rational soul, though only a part, be considered more perfect than the whole? Harvey avoids this conclusion by changing the rank order. Granted, the teleological order of the world is not bound to calculation and consciousness; on the contrary, God's efficacy in nature comes about unconsciously, "blindly," fatefully (*tanquam fato,* DG 195, W. 365). *But precisely this is the more perfect activity, since it needs no deliberation;* in it, means and end are not separated—material, efficient, and final cause are identified.

Human planning and art are the highest level that an individual being is capable of on its own, but what natural entities as divine instruments reveal in formative power and craftsmanship is still higher, an unattainable ideal. In his vital aspect, man, too, is himself only such an instrument. All his planning and all his art depend on the events in his organism, which are beyond his will, on the processes of conception and circulation, the "everyday miracle"; they cannot excel this miracle or even repeat it.

e. SUMMARY: CYCLE AND POLARITY

To discuss the meaning of the symbolism of the circle for Harvey's thought would be to carry coals to Newcastle—it is so clear in all his writings and has been so thoroughly investigated. However, these discussions seem to have permitted another point of view to escape the attention it deserves: the principle of *polarity*, which represents the essential supplement to that of the circle, and which has been shown to be significant at various junctures in our investigation. For the living is formed as much by *rhythm* as it is by cyclical processes, and in Harvey's view, rhythm is always the expression of a polar constellation.

Moreover, *polarity and reciprocity also underlie the various circular motions as condition and stimulus*. The antagonism of heart and blood brings about systole and diastole, and produces the cyclical movement of the organism. In the same way, it is the polarity of male and female that sustains the circle of "hen and egg," of egg and organism, and thus maintains the circular movement of the species. In both cases, polarity serves especially for the higher development and perfection of the organism or of the species.[175] Finally, the alternation of approach and withdrawal of the sun produces the cycle of life, generation, and corruption as such. But above all stands the polarity of the two aspects of nature itself: of a thing-like material aspect in the foreground, which is "transsubstantiated" into the higher, divine aspect.

> In this way all natural bodies fall to be considered under a twofold point of view, viz. either as they are specially regarded, and are comprehended within the limits of their own nature, or are viewed as instruments of some more noble agent and superior power . . . therein is it, that they seem to participate with another and more divine body, and to surpass the powers of the ordinary elements.[176]

All movements in nature arise from the union of the "lower" and "higher" world, interact harmoniously together, and form the order of the cosmos (334, W. 508).

Thus, if polarity lies at the basis of the cycles, it also shapes their very course. The points on the circle are not simply "equivalent" to one another, but constitute a *gradient*. This holds for the inclination of the ecliptic with the solstices as its "poles," and equally, for the "revolution" from egg to organism and back (DG 115, W. 285). Ontogenetically, the development from the egg proceeds upward to the mature animal, but seen ontologically, it goes downhill, namely, as the unfolding of a form already given at the start. Finally, we have seen in chapter II that the circulation of the blood is also distinguished by a haemodynamic as well as a qualitative gradient. Heart and periphery can be conceived as its two poles: the periphery is the place where the impetus of the heart has weakened to a minimum, but at which the centripetal tension of the blood is greatest (declinante *sponte sanguine . . . ad centrum*, DMC 22, W. 37; proclivis . . . *est . . . sanguis, ut a circumferentia moveatur in centrum*, 85, W. 71; author's ital.) On the other hand, the blood here loses its vital heat and is in danger of clotting. At the other pole, the heart is the place of the most forceful impulse, the location of the greatest heat, and the "resting point" of the blood.

Even from what we have said so far, an *asymmetry* is evident in the polarity. The two poles are not related to one another as equivalents like "north" and "south," but one is ranked before or above the other. In the alternation of coming to be and passing away, life is the primary condition of things; it cannot be derived from materially preformed constellations. The lifeless arises at all only with the death of the living: the stuff that has dropped out of the whole, as it were, broken apart into isolated parts or elements, and now subject to its own laws.[177] As to the polarity of male and female, Harvey is inclined to reverse the traditional rank order. The female, which according to the dominant doctrine provided only the material for the formative power of the semen, now produces, by a kind of parthenogenesis, an egg that stands in need of the sperm only as a supplemental condition for its further development.[178] We have already observed that the blood forms the basic element in its reciprocal relation with the heart; it contains in itself, at the start, the polarity of contraction and dilation, and then constructs the heart as its "impulse amplifier." The vital aspect produces the mechanical; and even later, through its vital heat, the expansion of the blood remains the trigger for cardiac action. The heart itself again broadens this elementary movement to the polar components of auricular and ventricular pulsation, which make possible a further widening of the radius of action.[179] As we can see, the asymmetry of polarity is the expression of its unfolding from a *primary unity*, which stands closer to one of its descendants than to another.

From this point of view, light is shed also on the question of the primacy of heart or blood in the organism. In the *De Motu Cordis*, the heart appears as ruling principle; in the *De Generatione*, the blood. The literature has interpreted this in various ways: as unresolved contradiction or as a development in Harvey's thought.[180] However, we must consider against the second view, on the one hand, the fact that in the *De Motu Cordis*, Harvey already describes the blood as *primum vivens ultimum moriens*, and on the other, that in the *De Generatione*, he still ascribes to the heart the royal attributes that now distinguish the blood as well: first seat of the soul and of vital heat, perpetual hearth of the living being, spring and source of all its powers (168, W. 341).

In a certain sense, there is something in both interpretations.

The contradictoriness as well as the development of Harvey's views in fact reside in the circumstances themselves. As *arché* of the organism, as a unity as yet undifferentiated, the blood controls the first stage of embryogenesis. With the increasing articulation of the embryo, the reciprocal interaction of blood and heart arises, the circulation unfolds. But now, there also emerges in the circulation a polarity, a gradient that makes the heart the higher pole of the motion of the blood in the mature organism so that the heart can now be seen as the starting point and end point of that motion. In his work, Harvey goes in the opposite direction, "inductively," from the differentiated phenomenon to the first and the simple—and he places the emphasis accordingly. We have followed him on this path. But the natural development proceeds "deductively" from principles that are simple in themselves to the differentiated and complex.

We can also describe this development as the unfolding of the polarity of *fluid* stuff on the one hand and on the other of the *solid* structures that proceed from it. The heart, like all the organs, is in its origin a congealed movement of the blood, as it were, and it remains dependent on the blood for its vitality as well as for its action. However, the heart increasingly imposes its control on the primary motion of the blood so that in the arterial branch of the circulation, the blood becomes the mere *object* of cardiac action. But on the whole, the polarity remains in equilibrium: conversely, the heart is also the "object" of the motion of the blood. Only in the isolated, excised but still beating heart is the complete independence of the "solid" from the "fluid" pole manifested.

Finally, let us look at polarity once more from the point of view of its *unity*. In the last analysis, the reciprocal relations in the organism stem from the elementary sensibility and mobility of the *punctum saliens*, that is, the vital principle of the blood. But developed polarity also remains a unity: as a totality of movement, for instance, in the case of the muscles, as rhythm in the heartbeat, and in respiration. Since polarity is thus already subordinate to a form and is in itself harmonious, it can in turn take its place in the service of a higher principle; it becomes the instrument of the will, the expression of the soul, or the foundation of human art.

Just as there is a unity at work here in what seems contradictory, so a universal principle controls the change of generations and the polarity of the sexes. It does not stand apart over everything, but "flows" in a kind of emanation through its instruments, proceeding from the parents through the sperm and the egg, from there to the *punctum saliens*, the blood, and the organism. The operative principle is contained *in* the phenomena; material, efficient, and formal cause are united in what Pagel has called "working matter": sperm, egg, *humidum primigenium*, and finally blood.[181] In place of the traditional hierarchy of ultimate and instrumental causes, we have a monistic and immanent principle, for which "spirit" and "heat" are mere synonyms. It is not specific, localisable "faculties," but the uniformly streaming totality of motion in the organism that explains all its processes. In this "systemic" vitalism, the organs are no longer controlling centers, but tools of the blood.

For Harvey, the defeat of the Galenic system lay in a renewed and, in many respects, radicalized Aristotelian vitalism. Yet the signs of the times pointed in another direction; and it was not, after all, Harvey himself who, through his reduction of complex hierarchies to a vital principle and its identification with the basic material of the organism, supported, even if unintentionally, a mechanistic development based on his discovery.

Notes to Part C

1. Numerous attempts have been made to demonstrate the knowledge, or at least a presentiment, of the circulation in the anatomical sense in European and non-European intellectual history *before* Harvey—without success as Pagel has once more shown (Pagel 1967, pp. 127–209).

2. Cf. E. Lesky, "Harvey und Aristoteles," in *Sudh. Arch..* 41 (1957), pp. 289–316, 349–378. The passage quoted is on p. 296. In fact, Bacon himself compared the category of teleology, central to Harvey's thought, with a virgin consecrated to God and therefore (scientifically) *barren* (*De dignitate et augmentis scientiarum*, III, 5. In *The Works of Lord Bacon*, London, 1841, Vol. II, p. 340). For further literature on this conception of Harvey's, see Pagel (1967), p. 21f.

3. E. Ackerknecht, *Kurze Geschichte der Medizin*. Stuttgart, 1959, p. 91.

4. A. Castiglioni, *Histoire de la Médicine*. Paris, 1931, p. 418.

5. T. Ballauf, *Die Wissenschaft vom Leben. Eine Geschichte der Biologie.* Freiburg, 1954. In my view, the last statement in particular is diametrically opposed to Harvey's view of living things.

6. G. K. Plochmann, "William Harvey and his Methods," *Studies in the Renaissance* 10 (1963), pp. 192–210.

7. H. Driesch, *Geschichte des Vitalismus*. Leipzig, 1923 (1st ed. 1905), pp. 23ff.

8. Pagel, 1967, p. 251ff.

9. C. Webster, "Harvey's *De Generatione*: Its Origins and Relevance to the Theory of Circulation," *Brit.J.Hist. Sci.* 3 (1967), pp. 262–74, p. 262.

10. Thus Garrison: "In endeavouring to locate the motor power of the muscle (i.e. of the heart; author) itself, in his attempts to explain the

functions of the blood and lungs, Harvey fell into a phase of medieval mysticism, derived from Aristotle's doctrine of the primacy of the heart, as the seat of the soul. . . ." (p. 247).

11. Needham, p. 149. According to the opinion of his translator Willis as well, Harvey was "possessed by scholastic ideas" in the composition of the *De Generatione* (*Works*, pref. p. lxxi). For the evaluation of Harvey's *De Motu Locali Animalium* (1627, edited and translated by G. Whitteridge, Cambridge, 1959; referred to hereafter as DML), see the introduction by G. Whitteridge, p. 4.

12. Pagel, 1976, p. 1.

13. ". . . nulla enim est scientia, quae non ex praeexistente oritur," in W. Harvey, *Exercitationes duae de circulatione sanguinis ad Johannem Riolanum filium*, 1649. *Opera*, Pt. I, pp. 105–67; Willis, pp. 87–141; referred to below as ER, p. 108; W. 89; p. 155; W. 123.

14. Aristotle, *Analytica posteriora* I,1, 71 a 1; referred to below as AP.

15. "Prae ceteris autem, Aristotelem ex antiquis . . . sequor," DG, preface (unnumbered, referred to henceforth as "Pf."); Willis, p. 167.

16. Aristotle, *Physics*, I,7, 190b20ff.

17. ". . . now there is contrariety between becoming and being—for what is posterior in becoming is prior according to nature, and the first is the last in coming to be (for the house is not for the sake of the bricks and stones, but these for the house)." Aristotle, *De Partibus Animalium*, II,1, 645a25ff. (cited hereafter as DPA). Just so Harvey: "Quod enim natura prius est, id fere in generationis ordine posterius existit" (DG 232; W.406). "Quicquid enim praestantius est, id quoque *natura prius* est; unde vero aliud producitur, id *tempore prius* reputandum est" (DG 114; W. 285; my italics).

18. "exposita, qua ego usus fuerim, indagandi methodo" (Pf.; p. 152). "Quamvis ad scientiam quamlibet via unica pateat, qua nempe a notioribus ad minus nota, et a manifestis ad obscuriorum notitiam progredimur; atque universalia nobis praecipue nota sint (ab universalibus enim ad particularia ratiocinando, oritus scientia); ipsa tamen universalium in intellectu comprehensio, a singularum in sensibus nostris perceptione exsurgit" (Pf.; W. p. 154).

19. DG Pf., W. p. 154ff. Although in essentials Harvey reports Aristotle appropriately, lacking the original he cannot, in my view, have understood the (admittedly difficult) passage in the *Physics* correctly (even though none of the commentators have come to the same conclusion: cf. Lesky, p. 290ff.; Plochmann, p. 197f.; Pagel 1967, p. 34ff.) In this case, the two sentences are not to be understood as complementary, but in fact—though this is exceptional—have the *same* meaning, that is, they both refer to *induction*. In the *Physics* passage as well as in the other, Aristotle is speaking of the necessity of proceeding from what is prior for *us* to what is prior *according to nature*; and the next sentence: "For this reason thought must proceed from the general to the particular" does not prescribe the deductive contrary of induction, but is explained in what follows: the sensibly perceptible phenomenon is still an *unanalyzed whole*, which has yet to be analyzed into its *particular determinants* or aspects (cf. also the commentary on the

Physics by H. Wagner, p. 395). But such a whole is in a certain sense a "universal" ("tò dè kathólou hólon tí estin," 184 a 26), that is, with respect to its *particular principles*.

That Harvey interpreted this misleading passage quite differently is clear not only from his juxtaposition of the two quotations, but also from the commentary he adds to it: here, he transfers the characteristics of the sensibly given, thus, according to Aristotle of the unanalyzed whole, to the universal in the sense of the premises and principles: it is the *latter* that is said to be a *"totum et indistinctum quid"* (Pf., W.). For Harvey, the sensible phenomenon seems also to stand on a higher level; it is *"clarius, perfectius,"* abstraction on the other hand *obscurius* (Pf., W. p. 157), less definite, and therefore, more susceptible of subjective interpretation. (In this connection, Harvey cites a passage in Seneca in which he explains the diversity of possible perspectives and artistic reproductions of the same object, and Harvey cites this as a parallel to science. Pf., W. p. 156.)

Thus Harvey (mis)understood the passage in the *Physics*, which contradicts his preference for the phenomena as against their interpretation, the empirical over against the rationalistic; in Aristotle, the two stand in a more balanced relation to one another.

20. W. Harvey, *Exercitatio anatomica de motu cordis et sanguinis in animalibus*, 1628, *Opera*, Pt. I, pp. 3–103; Willis, pp. 1–86 (referred to as DMC). On Aristotelian method in the DMC, cf. Plochmann, p. 200ff.

21. "Hinc causam aperte videbis, cur in Anatome tantum sanguinis reperatiur in venis, parum vero in arteriis." DMC, p. 62; W. p. 51.

22. Letter to R. Morison, April 28, 1652, Willis, p. 604.

23. "Quod si satis cognita habebuntur, tunc sensui magis quam rationi erit credendum. Rationi etiam fides adhibenda est; si, quae demonstratur, cum iis rebus conveniunt, quae sensu percipiuntur" (DG Pf.; W. p. 163). Cf. Aristotle, DGA III 10, 760 b 31 ff.

24. Lesky, pp. 301 ff.; e.g.: "ex quibus observatis rationi consentaneum" (DMC p. 26; W. p. 22); "sensui contrarium est et rationi" (DMC p. 60; W. p. 49); "confirmata omnia et rationibus et ocularibus experimentis" (DMC, p. 81; W. p.68); "ubi per autopsiam contraria, eaque rationi consentanea ipsemet (lector!) propriis oculis certior factus, deprehenderis" (DG Pf., W. p. 152), among others.

25. Kuhn, esp. pp.19, 23ff. ("The nature of normal science"). On Harvey, see e.g. G. Whitteridge, *William Harvey and the Circulation of the Blood*, London, 1971, p. xi: " . . . I believe that Harvey falls into the category of the great scientist who is not conscious of any philosophical method underlying his actions, yet by these actions advances knowledge." Similarly, K.D. Keele, *William Harvey, The man, the physician and the scientist*, London, 1965, p. 108: "It is probably because of the unsatisfactory nature of Harvey's own interpretation of his scientific methods that so little attention has been paid to his own account. Indeed, his own concept of scientific method was almost purely Aristotelian . . . The uniqueness of Harvey's achievement in the seventeenth century can perhaps be best measured by the gap between what he thought were his methods and what we, from our respective vantage point, can see them to have been."

26. Kuhn, p. 88.

27. ". . . sensibilia sunt per se, et priora; intelligibilia autem, posteriora, et ab illis orta"; ". . . ista citra experientiam . . . haud melius intellexerit, quam caecus natus, de colorum natura, et discrimine, aut surdus, de sonis iudicaverit"; DG Pf.; W. 157; cf. ER p. 154; W. 131: "They show the judgment of the blind in regard to colours, of the deaf in reference to concords."

28. Galileo Galilei, *Il Saggiatore*, 1623, *Opere*, ed. A. Favaro (Edizione Nationale), Florence 1890-1909, vol. 6, p. 232; quoted in Crombie, p. 374.

29. ". . . praesertim, cum tam apertus facilisque Naturae liber sit" (DG Pf.; W. 152). P. Sloan rightly speaks of the "epistemological incoherence between the two very prominent wings of the developing new science": Aristotelian realism is sharply opposed by the new physics, but once more acquires increasing significance for medicine and biology. "If one wished to press the point, the arguments of a Galileo could be turned against a Harvey, and vice versa" (*op.cit.*, p. 7).

30. "Astronomiae exemplar non hic imitandum est, ubi ex apparentis dumtaxat, et ipso quod sit, causae et ipsum propter quid investiganda veniunt" (ER 145; W. 124).

31. "*Foecundum* autem appello, quod (nisi alicunde impediatur) a vi efficiente insita ad destinatum finem pertinget; idque, cuius gratia instituitur, assequetur" (*DG* 187; W. 361). Compare the Aristotelian definition of end, for example, in *De Partibus Animalium* (= *DPA*): "We always say that something is for the sake of something when an end appears with respect to which the motion is completed if nothing interferes" (DPA I, 1, 641b 24 ff.).

32. ". . . finalis causa, tam in natura, quam arte, reliquarum omnium prima est . . . Inest enim quodammodo, in omni efficiente, ratio finis; a quo finis; a quo illud, cum providentia operans, movetur" (DG 401; W. 583).

33. ". . . the whole has a separate constitution and existence before its parts, the mixture before the elements" (". . . totum, suis partibus prius constituitur, et decernitur; mistum prius, quam elementa." DG 167; W. 340). "The eye sees, the ear hears, the brain perceives, the stomach digests, not because such characters and structures (naturally) belong to these organs; but they are endowed with such characters and structures to accomplish the functions appointed them by nature" (". . . oculus videt, auris audit, cerebrum sentit . . . not quia illarum partium talis temperies, et fabrica contigit; sed organa ista, ut destinatas a natura operationes obeant, eiusmodi temperie et fabrica donantur." DG 364; W. 541).

34. Pagel, 1967, p. 131. Although Pagel brings to the fore the previously neglected teleological aspect of Harvey's thought—perhaps even excessively—he counts it as belonging to Harvey's "unscientific" side (*ibid.*). Thus also in W. Pagel, "Harvey's Role in the History of Medicine," *Bull. Hist. Med.* 24 (1950), pp. 70–73: "In a contemporary scientist . . . non-scientific leanings cannot possibly as intimately be bound up with his scientific work as those of a seventeenth century savant" (*ibid.*, p. 73). Here the modern conception of science is dominant even in Pagel's case; Harvey would surely have rejected this division.

35. "Quo pacto igitur ars nobis advenit; eodem omnino cognitio et scientia acquiritur: nam ut ars circa facienda, ita scientia circa cognoscenda, est habitus" (DG Pf.; W. 156–57). W. Brunn also calls attention to this passage (*Die Kreislauffunktion in William Harvey's Schriften*, Berlin, 1967, p. 86). Harvey refers to a longer passage in Aristotle's *Posterior Analytics* (II, 19).

36. See note C 11 above, or G. Keynes, *The Life of William Harvey*, Oxford, 1966, p. 163: ". . . (Harvey) never made a final revision, though this is perhaps less regrettable in that the work contains little that can be called original." Only Brunn (p. 101) and J. Stannard, "Aristotelian influences and references in Harvey's *De Motu Locali Animalium*," in *Studies in philosophy and in the history of science: Essays in Honor of Max Fisch*, Lawrence, Kansas, 1970, pp. 122–31, come to a more positive judgment. They, too, gave only a partial assessment of the manuscript.

37. See for example Whitteridge, pp. 41ff., Pagel (1987), pp. 127ff., or Bylebyl, "The Medical Side of Harvey's Discovery: The Normal and the Abnormal," in Bylebyl, ed., *William Harvey and His Age*, Baltimore: Johns Hopkins University Press, 1979, pp. 28–102.

38. Embryological themes are included, for example, in the preface, chapters 4, 6, 16, and 17 (DMC 19f., W. 18; 34, W. 28f.; 43ff., W. 36ff.; 87f., W. 73; 93, W. 77f.; 98f., W. 82). It is striking that Harvey's application of animal experimentation to man has often been emphasized in the literature—while the same practice has happily been mentioned as a reproach in the case of Galen.

39. See, for example, Bylebyl, p. 76: ". . . (Harvey's) fundamental innovation was virtually identical with the consideration of the quantity of the blood"; or S. Peller, "Harvey's and Cesalpino's Role in the History of Medicine" in *Bull.Hist.Med.*, 23 (1949), pp. 213–35: "It was Harvey's quantitative reasoning which brought observations and experiments under one hat and produced a new theory" (p.229). "The spirit of measuring, weighing and counting now slowly permeated medicine, routine as well as research" (p.233).

40. See F.G. Kilgour, "William Harvey's Use of the Quantitative Method," *Yale Journal of Biology and Medicine* 26 (1954), pp. 410–21: "Apparently, quantitative evidence was not important in leading Harvey to develop the idea of circulation" (p. 419); or F.R. Jevons, "Harvey's Quantitative Method," *Bull.Hist.Med.* 36 (1962), pp. 462–67: "Harvey's quantitative experiments were thought-experiments" (p.465).

41. Cf. DMC 12, W. 11f. Lord Birkenhead considered this thought the real "spark" of the discovery (Birkenhead, "The Germ of an Idea or What put Harvey on the Scent?" *J.Hist. Med.* 12 (1957), pp. 102–5.

42. Boyle's report is given in Keele, p. 134.

43. Fleck (1980), pp. 35ff., English edition (1979), pp. 25 ff. See Part A above.

44. Pagel (1967), p. 110; cf. also pp. 103ff., 113ff.

45. ". . . (Harvey) was of course aware of the analogy that could be drawn between the movement of the blood and the various circular motions instanced by Aristotle. But this is not to say that Aristotle's philos-

ophy, any more than that of Paracelsus, has any part whatsoever in the formulation of the hypothesis. If this were so, one might as well ask why no-one before him had reached this conclusion" (Whitteridge, p. 128).

46. "Quo longius arteriae distant a corde, eo minore multo vi, ab ictu cordis per multum spatium refracto, percelluntur", and following; DMC 101, W.88. This corresponds to the late medieval *impetus theory*, according to which a (violent, non-natural) movement is impressed on the object, an "impetus" which is lost in the course of the movement. On the role of the impetus theory in Harvey, see Brunn, pp. 63ff. and Pagel (1976), pp. 67ff.

47. "... sanguis sponte sua versus principium (quasi pars ad totum, vel gutta aquae sparsae super tabulam ad massam) facile concentratur et coit (uti a levibus causis solet celerrime frigore, timore, horrore et huiusmodi causis aliis); praeterea, e venis capillaribus in parvas ramificiationes et inde in majores exprimitur motu membrorum et a musculorum compressione; proclivis etiam est et magis pronus sanguis, ut a circumferentia moveatur in centrum quam contra ... Unde sequitur; si principium reliquit, et loca stricta et frigidiora init, et contra spontaneum movetus sanguis, violentia opus habere et impulsore: tale cor solum est" (DMC 84f., W. 70f.).

48. "... and if there is to be violent movement, there must be natural movement, for violent movement is contrary to nature, and the unnatural is posterior to the natural" (Aristotle, *Phys.* IV 8, 215a1ff.).

49. There are a number of passages in Harvey's work which point to his still-Aristotelian understanding of physics: "In mundo majore, terra in centro posita, aqua et aere circundatur" (DG 44, W. 213); "nec magis *naturaliter gravia omnia ... deorsum tendunt*, aut *levia sursum* moventur; quam semen, et ovum, in plantam, aut animal ... feruntur" (DG 195, W. 369; author's italics).

50. "... inter hos duos motus tempus aliquod quietis intercedit; ut cor *quasi suscitatum* motui *respondere* videatur" (DMC 31, W. 26; author's italics). Cf. also, Harvey's London anatomical lectures (1616): "cor respondet auriculis ut quod in ipsum impulsum ipse propellat" (*Praelectiones Anatomiae Universalis. The Anatomical Lectures of William Harvey*, ed. and trans. G. Whitteridge, Edinburgh, 1964, p. 268).

51. "... post duas vel tres pulsationes auricularum ... quasi expergefactum cor respondere," etc.; DMC 32, W. 27). It is remarkable how, without electrophysiology, Harvey not only recognized the significance of the right auricle for cardiac action (this is even given special emphasis in chapter 17 as *primus motor*; DMC 97, W. 80), but, as Keele remarks (p. 129), also describes here the disturbance of atrioventricular transmission, from a 2:1-, 3:1-, up to a total atrioventricular block.

52. "... in ipso sanguine, qui in dextra auricula continetur, obscurum motum, et inundationem, ac palpitationem quendam manifesto superfuisse; tamdiu scilicet, quam calore et spiritu imbui videretur"; DMC 33f., W. 28.

53. "... ex qua ... fiunt cordis auriculae, quibus pulsantibus perpetuo inest vita"; DMC 34, W. 28. Cf. also ch. 17: "vesicula pulsans ... in auriculas ... transit; super quas cordis corpus pullulare incipit" (98f., W. 82).

54. "... nature in death ... retracing her steps, reverts to whence she had set out, returns at the end of her course to the goal whence she had started; ... whence that in animals, which was last created, fails first; and that which was first, fails last." (Ita Natura in morte, quasi decursione facta, reducem ... agit, motu retrogrado ... eo unde proruit ... ; unde, quod in animalibus ultimo fit, deficit primum et quod primo ultimum; 34f., W. 29). The parallel passage in Aristotle is DGA II,5, 741b19ff.

55. This terminology comes from Brunn, p. 56.

56. This passage also sheds light on the comparison of the expulsion of blood with the working of a *pump*, which is to be found only in a later addition to the *Praelectiones Anatomicae* (p. 272f.) and in the second polemic against Riolan (ER 159, W. 135). A great to-do has been made in the literature about the pump analogy since Harvey here seems to be anticipating the modern view of the heart; often, it is even seen as the foundation for his discovery. Nevertheless, the analogy, which he came upon only later, is in no way an integral component of his theory. This is clear from the very fact that, by rejecting *attraction*, Harvey is excluding precisely the *suction* characteristic of a pump, and instead, as we shall see, explains the filling of the heart through the *flow of blood*. Thus, the pump analogy illustrates only one component of cardiac action, the expulsion of blood. Cf. also G. Whitteridge, p. 169: "It seems likely that it was during these years [i.e. about 1635: auth.] that Harvey saw the analogy ... (it) does not occur in *De Motu Cordis* and it is unlikely that it formed any part of the original theory. ... Unlike that of Descartes, ... (Harvey's hypothesis) was not developed with reference to any mechanical system. ..." Similarly, Pagel (1967), pp. 212ff. or H.B. Burchell, "Mechanical and Hydraulic Analogies in Harvey's Discovery of the Circulation," *J.Hist.Med.* 26 (1981), pp. 260–77: "Harvey's references to mechanical analogues are cursory and never developed into a detailed comparison. His aim in such passages appears to have been illustrative ... Harvey's vitalistic philosophy ... might block any thoughts that would equate the heart directly with a mechanical pump" (p. 262). Cf. also G. Basalla, "William Harvey and the Heart as a Pump," *Bull.Hist.Med.* 26 (1962), pp. 467–70.

57. "Mechanica: sicut illud quod superat ea a quibus Natura superatur, et succurit difficultatibus cum quod praeter Naturam utilitatem fit" (DML 126). Here, Harvey refers to the pseudo-Aristotelian *Mechanica* (847a1ff.), where we read: "Astonishment is aroused ... by what results contrary to nature and through art (*téchne*) for the benefit of human beings. For at many points nature effects the contrary of what is useful to us. ... If we need to do something against nature, because of its difficulty, we are faced with a problem, and require art. Hence we call that part of art which solves such difficulties, mechanics."

58. "Sic in musculis Natura nusquam difficultatibus huiusmodi succurere deficit. Unde ... et hic tot vere miranda quomodo musculi vires ossa movent et annexa pondera" (DML 126).

59. DMA 7, 701b11ff. On *alloíosis* as qualitative change, see Aristotle, *Phys.* V,2, 226a27ff.

60. "... nullum motum sine musculo, [velut] instestinorum, cordis, pupillae" (DML 114). "An sic musculi in automatis; in actu motus est dum maiores et minores fiunt vicissim; et sic relaxatio pars actionis [est] ut cordis sistole diastole" (DML 152).

61. "Natura movet by signes et watch words parvo momento factis, [partes] circa partes principes. Aristoteles as shooting momentum in principio magnus error fine" (DML 146).

62. The adjacent comparison of cardiac motion with the *act of swallowing* points in the same direction (DMC 38, W. 32); if it represents a "gulping down" of the blood, it could also be triggered by the blood, as, in the mouth, the motion of swallowing is triggered by the food irritating the palate. Moreover, in the tenth chapter Harvey describes how, after experimental pinching of the vena cava, and thus *without admission of blood*, the heartbeat slows down and the heart "seems ... about to die" (*ob defectum sanguinis ... languidius tamen pulsare, sic ut emori denique videatur*, DMC 65f, W. 54).

63. DMC 84, W. 70. The Aristotelian passages on the *calor* doctrine on which Harvey relies are numerous and do not entirely agree among themselves. At any rate, heat is the presupposition for all life, and its loss is synonymous with corruption and death (e.g. *Meteorologica* IV,1, 379a2ff.; *De Juv.* 469b1ff.). In higher animals, its source is the heart (DPA III,6, 670a23ff.), on which therefore all tissues and organs depend. Heat is the "instrument" of the soul (DGA II,4, 740b25ff.); its tasks are threefold: (1) processes of qualitative change (*pépseis, concoctiones*; DPA II,3, 650a2ff.; *Phys.* VIII,7, 260a30ff.); (2) movement (DPA II,7, 652b9ff.; *DMA* 703b9ff.); (3) sensation and perception (DA III,1, 425a5ff.). To be sure, the *pneuma* also plays an essential role in the second and third tasks. See below. Cf. also Mendelsohn, pp. 11ff.

64. Brunn calls attention to this in his extensive investigations of the production and mode of operation of innate heat in Harvey (pp. 5ff., 22ff., 30ff.): a fire in the left ventricle "... would have to be extinguished by the abundance of blood streaming in without hindrance, or it would have come to an explosion" (p. 25).

65. See Brunn, p. 111 (note 7), with numerous references.

66. Thus too Aristotle: heat prevents the coagulation of the blood (DPA II,9, 654b10f.); the heart "... is for the blood the cause of its motion and heat" (*ibid.* III,5, 667b27f.). Cf. on the other hand Brunn (p. 32): "Blood produces and transports calor innatus through motion," wherewith Harvey's concept of *calor* would assume the character of a material, an interpretation to which I cannot subscribe. The passages cited by Brunn (pp. 195ff.) all speak either of *fire* (DG 332, W. 506, DMC 37f., W. 31), or of heat, insofar as it *distributes itself with the blood* (DMC 37f., W. 11; 69, W. 56; 82, W. 68; 86, W. 72). Naturally this is "material," insofar as it is bound to the blood, but not because Harvey "takes inner heat for a material" (Brunn, p. 107). This can indeed be alleged of Aristotle, who connects heat closely with the (subtle) *pneuma*. But for Harvey, the *spiritus* is precisely not separable from, but *identical* with, the blood (ER 137ff., W. 117ff.). Heat exists only in connection with something else (ER 139, W. 119), not, however, as *spirit*, but in the

form of *blood,* it streams through the body (ER 140, W. 120); "solus nempe sanguis est calidum innatum" (DG 328, W. 502).

67. Pagel frequently points this out, e.g. (1976), p. 15.

68. Harvey was thoroughly familiar with the work of Cesalpino; for a detailed treatment, see Pagel (1967), pp. 188ff.

69. "Quem motum (sc. sanguinis, auth.) circularem eo pacto nominare licet quo Aristoteles aerem et pluviam circularem superiorum motum aemulari dixit"; DMC 56, W. 46.

70. ". . . in vena et arteria coronali privato usui . . . sanguinem continetur"; DMC 84, W. 70. Even more clearly later in ER 162, W. 137f.: "ideo cordi arteriae et venae coronales assignantur . . . ad caloris influxum, pro fotu et conservatione ipsius . . . Hoc modo autumo, *calorem nativum . . . pulsus quoque efficiens primum esse*" (author's ital.).

71. DMC 84, W. 40. Compare here also Aristotle: to assume the production of heat from the air is absurd, since the respired air is colder than that exhaled (*De Resp.* 472b33ff., 473a10f.); that air serves rather for the cooling of cardiac heat (*De Juv.* 469b11f., 470a5) is shown also in fishes, which use water instead of air for this purpose (*De Resp.* 471b17ff.)

72. "Sanguis et spiritus unum corpus constituant" (DMC 12, W. 12), "sanguis vel spiritus" (34, W. 28), "tam calorem quam spiritus" (83, W. 69). etc. Cf. the *Praelectiones* 281f.: "spiritus et sanguis una res[,] ut serum et cremor in lacte[,] et [Aristoteles] "ratio sanguinis ut aqua calida", i.e., the concept "blood" is analogous to that of "hot water"—for according to Aristotle, the blood is *hot per se:* cf. DPA II,3, 649b22ff. Further: "Ut candela lux ita spiritus sanguine" (as light exists through the candle, so spirit through blood; *Praelectiones* 282). There follow some arguments for the thesis "*spiritus non ex aere*", among others Aristotle's example of the fishes. (See note 71.)

73. "circulari motu . . . in perpetuo motus" (DMC 81, W. 68); "perpetuo inest vita" (34, W. 28). Cf. also the later addition to the *Praelectiones*, 271f.: ". . . perpetuum sanguinis motum fieri."

74. Jean Riolan (1580–1657), *Encheiridium anatomicum et pathologicum*, Leyden, 1649; quoted in Whitteridge, p. 180f.

75. *Ibid.*, p. 182. As an objection to Harvey's disregard of the vital process in the *periphery,* Riolan's criticism seems not unjustified even from a modern point of view.

76. "Thus asserting in chapter eight that the heart would tend to drain the veins and overfill the arteries unless there were a return of blood from arteries into veins, Harvey was in effect using the familiar terms of humoral medicine to describe a constant quasi-pathological process caused by the heart-beat that is continually counteracted by a quasitherapeutic process at the periphery (i.e. through "phlebotomy within the body", auth.), with the two adding up to the new physiological principle of circulation" (Bylebyl, p. 81). For the same reason, Harvey had repeatedly to deal with the objection that his experiments (opening of veins, ligatures) were *interferences* with the organism, which therefore could say nothing about its physiology, but only something about pathological alterations.

77. In this connection, Harvey still speaks occasionally of a *vis pul-*

sifica (DMC 70, W. 57) or of a *vitalis facultas et pulsus* (ER 160, W. 136), but in the sense of an autonomous, vital motive force of the organ.

78. "impleatur tantum pars", DMC, 71, W. 58. In ER 150ff., W. 128f., Harvey begins to see this in a more differentiated way: the circulation is modified by local factors like the permeability of tissues, the blood supply to organs, the temperature of the extremities, viscosity of the blood, and local changes in the blood supply influenced by emotion.

79. See Harvey's letter to Morison, April 28, 1652, W. 604–610 (esp. p. 605). On this theme cf. also ER p. 165f., W. 140: here, Harvey criticizes the conception according to which the *vegetative faculty* in plants and animals works in the same way and thus the local *attraction* in animals as well renders a central impulse superfluous. The difference between plants and animals consists, Harvey declares, precisely in the special need of animal organisms for heat, a need which can be satisfied only through that impulse and the *circulation*.

80. "Prius in confesso esse debet, quod sit circulatio, antequam propter quid fiat, inquirendum; name ex iis, qua in circulationes et hac posita obveniunt, usus utilitates, investigandae sunt" (ER 165f., W. 122f.). Similarly, in the letter to C.Hoffmann of May 20, 1636, complete English translation in Whitteridge, pp. 248ff.

81. E.g., by L. King, *The Growth of Medical Thought*, Chicago: University of Chicago Press, p. 152: "In other words (!), *what* happens is more important than *why* or even *how* it happens."

82. Kuhn, 1962, p. 55.

83. ". . . quo magis motus onerosus eo magis organum corpulentum est et robustum ratione mechanica, unde aliis animalibus spiritus tandtum[,] alliis fibra, caro, etc." (DML 96).

84. "spiritus . . . maxime alterabilis a caliditate et frigidittate," DML 98; similarly, 24, p. 96. Cf. Aristotle DMA 7, 701b15f. and 10, 703a20ff.

85. ". . . for the capacity to perceive is located there [sc. in the center], so that, when the place around the origin [of movement] . . . is altered, this also affects the adjacent parts, which stretch and contract, so that the movement of animals necessarily originates in this way" (DMA 9, 702b20ff.). Harvey, too, quotes this passage (DML 96).

86. "Dubium. An? motus fiat repletione et distendendo . . . ebullitione, Aristotelis de cordis . . . An motus fiat inanitione, contractione . . ."; DML 100.

87. ". . . spiritum per arterias influere motum ex corde; sanguis et spiritus una res, et musculus et spiritus motivus, unde nutriri spiritus ut corpus; DML 102.

88. " . . . itaque . . . sensus motus gratia et motus a sunsu non solum quando et ubi sed qualis et quomodo", p. 104; "sentimus nos movere et absque hoc non fiunt (motus animales)", p. 42.

89. "Nervi itaque rivorum in morem a cerebro seu ex quodam fonte deducunt musculis facultates"; C. Galenus, *De Motu Musculorum* I, 1; in *Opera* (Kuhn), IV, p. 371.

90. ". . . sensitiva et motiva potentia videntur a corde [esse]. . . .

Actum [fieri] a sensibile et non a cerebro" (DML 108). As to the interpretation of the somewhat cryptic second sentence: "act" means on the one hand the *activity*, the *actual movement;* on the other, the *actualization* of the "potency" mentioned in the previous sentence. In the sentences that follow, there is talk of the mobility of the muscles independently of the nerves; this attests to the fact that it is the muscles that are being referred to as the "sensitive." Similarly, on the next page: "musculi ... cum *in actu* quasi palpitant" (p. 110). If, however, "sensibile" were referred back to "corde," this would not alter the fact that the heart transfers the "capacity for sensation and motion" to the muscles.

91. "... vegetativae motus licet cum organo ut ... cordis, intestinorum, uteri, *nec per nervos*" (DML 108). "Sunt tamen quae a sensu dependent tamen non in nostra potestate: (motus) intestinarum, matricis, vomitus, urinae ... [also later] cordis" (p. 40).

92. Galen, *De Usu Partium* V, *De Naturalibus Facultatibus* III; cf. O. Temkin, "The Classical Roots of Glisson's Doctrine of Irritability," *Bull.Hist.Med.* 38 (1964), pp. 297–328, esp. 306ff.

93. "... motus alii conveniens est semper fieri nec modulari pro arbitrio ut cordis, intestinorum"; DML 110.

94. "alii (motus) modulari (conveniens est) ... ; iis opus nervis cuius interventione sensus [et cuius] communicatione cum cerebro intellectus"; DML 110.

95. "Motus enim proprius sensibile quoddam.... Nervi usus communicare sensibile cerebro ut fiat iudicium ... sic nervis scissis nervosus [sed] inutiles motus"; p. 110. G. Whitteridge translates "motus enim proprius sensibile quoddam" as "Movement is then the characteristic of that which is itself sensitive" (p. 111). In my view, this cannot be correct: "proprius" modifies "motus"; it is not the scholastic "proprium" ("characteristic"). Harvey is using "sensibile" here in its second sense ("sensible", perceptible, in contrast to "actum a sensibile," i.e. "of that which is *capable* of sensation", cf. n. 87.) This interpretation follows also from the next sentence quoted in the text: *communicare sensibile cerebro.*

96. This comparison is also made by Aristotle, DMA 10, 703a30ff.

97. "An vicissitudo ut ferrum equilibrium. An compas to the lodestone. An per motum ebullitionis et palpitationis flammae et motus cordis, Aristoteles. An sic musculi ... maiores et minores fiunt vicissim; et sic relaxatio pars actionis ut cordis sistole diastole"; DML 152.

98. Observations of this kind are to be found, for example, in Chapter 17: the *punctum saliens* reacts to touch, pricking, change of temperature, and other irritations with movements or with change of frequency of its pulsations (*ad quemlibet vel minimum tactum, videbis punctum hoc varie commoveri, et quasi irritari ... Vidi ... ab acu, styli, aut digiti contactu, imo vero a calore aut frigore vehementiori admoto ... punctum hoc varia sensus indicia, pulsuum nempe varias permutationes ... edidisse*, DG 70, W. 239). Similarly in Chapter 52: from its various reactions, one can infer that the blood can perceive harmful and advantageous influences (*ex vario ipsius motu, in celeritate aut tarditate ... eum et irritantis initiam, et foventis commodum persentiscere*; 205, W. 380).

99. "Motus . . . *naturales* . . . a cerebro non dependent . . . tamen prorsus citra sensum non fiunt . . . utpote a qua (sc. sensu, auth.) excitentur, irritentur, et permutentur" (DG 256, W. 430f.).

100. "Quicquid enim sensus plane expers est, non videtur ullo modo irritari . . . posse. Nec certe alio indicio, animatum . . . a mortuo . . . internoscimus, quam per motum a re aliqua irritante excitatum" (DG 258, W. 432). As early as the *Praelectiones*, Harvey states that the body contains no parts without sensation and none without life (p. 6). But only with the developed conceptual apparatus of the *De Generatione* does Harvey anticipate the later "irritability" of Glisson and Haller. On this theme, see Temkin (1964) as well as Pagel, "Harvey and Glisson on Irritability. With a Note on Van Helmont," *Bull.Hist.Med.* 41 (1967), pp. 497–514.

101. ". . . dari sensum quendam tactus qui non referatur ad sensum communem . . . ac propterea in eiusmodi sensu, non percipimus nos sentire . . . quem (sensum) propterea a sensu animali distinguimus" (DG 258f., W. 432f.). In this context, "touch" is not to be taken in the narrow sense: according to Aristotle, it is directed to ". . . many different kinds [of objects], and these have different sorts of oppositions, such as hot-cold, dry-moist and others of that kind" (DPA II,1, 647a17ff.). In this sense, Aristotle also calls touch the primary sense without which no others can exist; for the activity of the other sense organs is also a kind of touching of their respective proper objects (DA II,2, 413b5ff. and II,13, 435a10ff.). In this context, Aristotle relates touch especially to the heart as primary sense organ (*De Juv.* III, 469a10ff.).

102. "The muscles . . . when affected with spasms and convulsions from some irritating cause, are assuredly moved no otherwise than the decapitated cock or hen, which is agitated with many convulsive movements . . . but all confused and without a purpose, because the controlling power of the brain has been taken away: common sensation has disappeared, under the controlling influence of which these motions were formerly coordinated, [with rhythm and harmony,] to progression by walking or to flight" (*Certe musculi . . . in spasmo et convulsionibus, a caussa aliqua irritantur non aliter moventur, quam gallus . . . detruncato protinu capite, multis . . . motibus agitatur, sed confusis omnino et irritis; quoniam potestas cerebri ablata est, et sensus communis evanuit, cuius antea moderamine, cum rhythmo et harmonia, motus illi ad progressum, aut volatum regulabantur*, DG 259f., W. 433).

103. ". . . pulsus ex eo [sc. sanguine] ortum ducat. Cum enim duae sint pulsationis partes . . . horumque motuum distentio prior sit: manifestum est, actionem illam sanguine competere" (DG 201, W. 375).

104. "Fit, inquam, diastole a sanguine ab interno quasi spiritu intumescente; adeoque Aristotelis sententia de pulsatione cordis (*fieri eam* scil. *ad modum ebullitionis*) aliquatenus vera est" (DG 201, W. 375).

105. "Certumque est, vesiculam dictam, ut et cordis auriculam postea, (unde pulsatio primum incipit) a distendente sanguine, ad constrictionis motum irritari" (DG 201, W. 375). Here, Harvey is referring to the *right* auricle, the first mover of the blood (DMC 97, W. 80), the starting point and end of the blood stream (ER 159, W. 135), where the vital power and

the pulse have their origin (ER 160, W. 156). The *left* auricle (and consequently also the left ventricle) acts in synchrony and harmony with the right (ER 156, W. 133).

106. Cf. DMC 97, W. 80, where this impulse strengthening is compared to a ball player's stretching back his arm for the throw; ER 156, W. 132f; also Harvey's letter to Morison (4, 28, 1632): ". . . the heart is expanded by the blood, which is thrown from the auricles into the ventricles; the ventricles are stimulated to contraction by this filling and expansion, and this motion always precedes the systole" (Willis 604; our translation).

107. ". . . distentionem etiam cordis motum quendam violentum esse, ad impulsionem . . . factum" (ER 165, W. 140).

108. ". . . pulsus itaque a duplici agente peragitur . . . haec mutua operae societate alternatum instituta, sanguis per totum corpus impellitur, vitaque inde animalibus perpetuatur" (DG 201, W. 375).

109. Possibly—and paradoxically—it was precisely Harvey's preference for the blood in contrast to the Aristotelian primacy of the heart that led him to adopt Aristotle's theory of ebullition, which he had already examined in the *De Motu Locali*. In consequence of Harvey's own exposition, a "slight signal" such as the streaming of blood into the auricle should have been quite sufficient to trigger cardiac action. But apparently, the polarity of the *vicissitudo* and the active role of the blood did not appear to him to be adequately represented by such a signal.

110. DG 201, W. 375; ER 156, W. 132, 161, W. 137. Since ancient times, fermentation has been the paradigm for organic expansion through inner heat; cf. also, DGA III,4 755a17ff., where Aristotle compares the growth of the egg with fermentation, but attributes *intraorganic* "fermentation" to the heat of the soul (*psýchikon thermón*). In the same way, Harvey insists that the usual picture of milk boiling over to represent the swelling blood can only be an analogy; "what happens to fluids *per accidens*, through an external cause, is effected in the blood through its own heat, or through an innate spirit" (*quodque in illis per accidens, ab agente externo [calore scil. alicunde adventitio] contingit; id in sanguine ab interno calore, sive spiritu innato efficitur* (DG 201, W. 375; also ER 161, W. 137). For blood is *essentially* warm (cf. notes 66 and 72).

For this reason, I cannot agree with Brunn's interpretation, careful though it is in other respects, when he describes the expansion of the blood as the effect of *cardiac heat*, which is produced by the coronary vessels (Brunn, 34ff., 38ff., 41). Nor do we see anywhere in Harvey's writings, how during its expansion "heat streams into the blood" (*ibid.*, 34). Brunn falls victim here to his own interpretation of heat as material (cf. note C-66). As we have just seen, Harvey guards himself explicitly against physical interpretations of the *calor innatus* and of the expansion of the blood; the heart is precisely not an "oven" or "heated pot" (*Neque cor [ut aliqui putant] tanquam anthrax, focus (instar lebetis calidi) caloris origo est et sanguinis*; ER 162, W. 137). The coronary vessels do indeed serve to heat it, but for its own maintenance and vivification, not in order to pass on the heat to the blood (ER 162, W. 137f.; see note C-70 above). The special heat of the heart is based simply on the quantity of the blood contained in it (*ideo calidiores omnes partes, quo magis sanguineae, et quo sanguinis magis abundant, calidiores convertibiliter dicuntur*, [ER *ibid.*]; also in the *Praelectiones*, 262). In the vena cava, where "it is collected in the greatest mass," and thus *already before it reaches the right auricle*, the blood increases in temperature through its "inner heat" (*ab interno suo calore incalescens;* ER 156, W. 132).

This later interpretation of *calor innatus* does not, indeed, contradict the conception of the *De Motu Cordis*: *calor > motus > calor*; however, Harvey now emphasizes the *inner heat formation* of the blood as against the aspect of heating through local motion.

111. ". . . ut illa elevatio, aut depressio sanguinis, non . . . sit causata, ab externo agente, sed ab interno principio, regulante natura" (ER 161f, W. 137; Willis has "an inner principle under the control of nature"). Cf. "id . . . secundum naturam ab anima regulatur" (DG 201, W. 375).

112. Cf. among other passages DMC 60f., W. 50, 83f., W. 70 as well as ER 150f, W. 128 (see above, note C-78).

113. On the question of the regulation of the circulation, see Brunn's exhaustive investigations (above all, pp. 34ff.), albeit with the qualification noted above; Brunn, too, relies only on the passages cited, above all "*regulante natura*."

114. "Natura in animalibus opera ope musculorum per rithmum et harmoniam facit et assequitur. . . . Divine patebit quod in celo dilectabile et amabile assequi motus harmonia rithmo cuius non habemus sensum non magis quam canes musicae" (DML 142).

115. "sensus non sine calore" (DML 92), "calidate ad sensum" (p. 88); "ille (sc. motus, auth.) a sensu; vel saltem concentratio et expansio a sensu et calore" (p. 24).

116. If we place this passage alongside Harvey's hymn-like encomium of divine harmony and rhythm, we might think that perfect, divine movement arises *through pure dissolution in its goal*; or, in Harvey's image, that the choir leader no longer directs with instrument or gesture, but through his pure inner anticipation of the sounds.

117. ". . . tanquam animal quoddam internum" (DMC 100, W. 83). Aristotle also sees the heart as the center of perception (above all, of *touch*, see notes B13 and C101) and as the source of the soul's sensations (DPA III,4, 666a11ff.). The expression "independent organism" occurs in many places (e.g. DPA III,4, 666a2, b17).

118. ". . . dubitandum non sit, quin punctum hoc [animalis instar] vivat, moveatur et sentiat" (DG 70, W. 239).

119. ". . . sanguine ab interno quasi spirity intumescente" (DG 201, W. 375).

120. "a spiritu fit pulsus cordis," "a spiritu caro ipsa contrahitur" (DML 94); "sanguis et spiritus una res, et musculus et spiritus motivus" (DML 102); "immediatum organum motivum . . . in quo spiritus motivus primo inest, sit contractile" (DMC 97, W. 81).

121. ". . . in musicis sonorum harmonia unde etiam mundo antipathia, sympathia," DML 142; "sic ab ordine et sympathia et antipathia actiones et passiones," DML 144.

122. DMC 88f., W. 74. The *De Generatione* (1652) is really Harvey's life work; the collection of materials goes back to the time of his London lectures (1618); large parts were already written in the 1630s, and the final version followed essentially in 1647–48. Cf. Whitteridge, p. 201 as well as Webster (1967).

123. DGA II,6, 742a33–b3.

124. Cf. note C114. In DGA II,4, 740a and II,5, 741b, there is already extensive exposition of the heart as *arché*.

125. "Quemadmodum enim pullum ex ovo nasci diximus, eodem omnino modo, atque ordine, hominis, aliorumque animalium generatio, contingit"; DG 85, W. 254. A "biologistic" position is evident in this emphatic statement, a position in which Harvey found himself in opposition to the more "anthropocentric" tradition; cf. E. Gasking, *Investigations into Generation, 1651–1828*, London, 1968, p. 27: "He is . . . arguing that all generation, including mammalian and human, should be viewed as special cases of the most general type exemplified by the hen. This is the exact reverse of all the previous theories, in which oviparous generation was understood in terms of the supposedly more typical and perfect viviparous generation."

126. Harvey was the first to recognize the significance of the blastoderm; cf. Lesky, p. 352, Needham, p. 136ff. The *colliquamentum* is the fluid of the blastula, the later amniotic fluid (Lesky, *ibid.*).

127. "ovum esse primordium commune omnibus animalibus" (DG 282, W. 456); "tale primordium in animalibus . . . est humor in tunica aliqua, aut putamine conclusus" (376, W. 554); "De Humido Primigenio" (339, W. 513).

128. DG 200, W. 374; 68, W. 237. Cf. Aristotle: ". . . the heart shows itself as a bloody point in the egg white. This point pulsates and moves, as if it were alive"; *Hist. An.* VI,3, 561a12f. According to Aristotle, the blood originates only after and *in the heart* (DPA III,4, 666a7ff., b24f.). Harvey contradicts this conception as early as the *Praelectiones* (p. 250).

129. ". . . fit sanguis, antequam punctum saliens efformatur; idemque calore vitali praeditus est, priusquam per pulsum cietur . . . sanguinem id esse, in quo quandiu calor vitalis non prorsus evanuit, potentia redeundi in vitam continuatur" (DG 200, W. 374).

130. ". . . quo tempore in ovo, de vita plantae, ad animalis vitam transitus fit (DG 66, W. 235); ". . . ovum anima vegetativa pridem imbutum, iam motiva, et sensitiva potentia insuper donari; et a planta in animal transiisse" (69, W. 239). In contrast, many lower animals (molluscs and insects, among others) fall back in the winter into a motionless, plantlike, vegetative stage: namely, with the exhaustion of the pulsation of their heart-like organs (*cum hieme latent et quasi mortua occultantur, vel plantae vitam tantummodo agant*; DMC 92, W. 76. Cf. Brunn, 45.)

131. Cf. DG 79, W. 248f.; 85, W. 254 (*eiusque parenchyma ex arteriis [unde materia affunditur] procreari*); 85, W. 255 (*parenchyma aliquanto post sanguinem nascitur, et vesiculis pulsantibus superadditur*); 204, W. 378. See also, Aristotle, DPA III,5, 668a27: the blood is "potentially body . . . and flesh. . . ."

132. "cunctarum . . . operationum vegetabilium instrumentum

commune, calor internus" (155, W. 327); "sanguis . . . simul . . . pars pulli videtur, et efficiens" (*ibid.*); "punctum pulsans, atque sanginem . . . sibi reliquum corporis, aliaque omnia pulli membra asciscere" (*ibid.*).

133. This sequence emerges from p. 68f., W. 238, as well as from 200, W. 374 (*fibras, et venas, posteaque vesiculam, et demum cor* . . .). When exactly the central impulse begins is not unequivocally clear from Harvey's account. His conception of the self-pulsation of the blood before the formation of all "receptacles" is unambiguous (*fit sanguis, . . . priusquam per pulsum cietur; atque . . . ab illo pulsatio incipit*; 200, W. 374; also 202, W. 376, or DMC 34, W. 28). But does the initial pulsation of the blood also become locomotion? The origin of the vessels after the appearance of the pulse, but *before* the formation of the contractile vesicle [i.e. of the heart] (DG 68f., W. 238); 200, W. 374) seems to necessitate this conclusion: since the vessels originate from the blood (85f., W. 255; 200, W. 374), the blood must have moved with its own momentum out of the *punctum saliens*. Granted, Harvey never explicitly mentions that kind of autonomous movement; in addition, on p. 66, W. 235f. there is talk of the systole and of assimilation and discharge of blood, even of the *punctum saliens*. As we said, it is not possible to determine when, according to Harvey's account, the *mechanical aspect* of the motion of the blood begins. He seems to have seen no discontinuity here: the contractile movement is derived immediately from the pulsation of the blood, the mechanical from the vital. Cardiac action is not something coming from the outside; it is conceivable only as unified with cardiac *content*. However, when Brunn speaks of an independent locomotion of the blood up to the time of the formation of the auricles (43), and even of a "natural circular movement of the blood needing no stimulus" (52), this interpretation cannot be reconciled with Harvey's account, nor brought into harmony, e.g., with the contraction of the vesicle (DG 69, W. 238) which already moves the blood mechanically.

134. "punctorum salientium alterum (adaucto foetu) in auriculas, alterum in ventriculos cordis abire" (DG 72, W. 241); "(cordis) parenchyma postmodum, in formatione foetus, vesiculae superioritur" (218, W. 392; also 85, W. 255); cf. DMC 98f., W. 84: ". . . ista vesica carnosior et robustior facta in auriculas . . . transit."

135. DG 168, W. 341; 85, W. 255. Harvey describes the further course of embryonic development in chapters 18, 65, 69, and 70; it is of little significance here.

136. ". . . ex sanguis corporis moles exsurgit . . . inde autem partes per divisionem obscuram delineantur primo, posteaque organa fiunt, et distinguuntur"; DG 166, W. 339.

137. "Pars, inquam in qua tum principium unde motus, tum finis sit; idem, *pater, et rex*" (DG 241, W. 415).

138. ". . . qui hoc modo philosophantes materialem duntaxat caussam assignant, et vel ex elementis sponte aut casu concurrentibus, vel ex atomis varie dispositis, causas rerum naturalium deducunt"; DG 38, W. 207. This was the case, for example, of Pierre Gassendi (1592-1695), Daniel Sennert (1572- 1637), or Nathaniel Highmore (1593-1684). But the Galenic

embryology revived by Fernel also adopted a preformatist point of view, according to which particular parts of the organism were already prefigured and contradicted the Aristotelian theory of embryology. Cf. Pagel (1967), pp. 237ff., (1976), pp. 23ff., Gasking, pp. 20ff.

139. "Quasi generatio nil aliud foret quam separatio, aut segregatio, aut dispositio rerum" (DG 39, W. 207); "naturae numen (quod summa arte, providentia et sapientia operatur . . .) non agnoscunt" (*ibid.*)

140. "Quemadmodum ex eadem pluvia incrementum capiunt omnia generis plantae; quia aqua, qui potentia erat similis omnibus, fit iam actu similis, postquam in earum substantiam transmutata fuit. Tum vero amarescit in ruta, acrescit in sinapi, dulcescit in glycyrrhiza, et similiter in reliquis"; DG 235f., W. 410.

141. DG 162f., W. 334f. The term comes from Harvey, though not what it designates: Aristotle, Peter Severinus (1542–1602), and Marcus Marci (1595–1667) were "epigeneticists" (cf. Pagel, 1967, pp. 235ff., 285ff.). Harvey also knows the opposite kind of process, the *sudden* and *accidental* union of preformed material as it can occur, for example, in the "spontaneous generation" of primitive organisms; he calls this "metamorphosis." Here, *chance and material* are the first causes; however, they produce only imperfect living things, which are above all *incapable of reproduction*, and even these they produce only out of already preformed organic material (164f., W. 334f.). The more perfect *blooded* animals, on the other hand, cannot arise in this way: epigenesis is characteristic for the gradual formation of the organism from the blood, and only this is *proprie dicta generatio*: generation properly speaking (162, W. 334).

142. "Quemadmodum igitur opera . . . ab arte perficiuntur; . . . cum figulus . . . imaginem ex luto, addendo, sive augendo, et figurando format; simulque materiam parat, praeparat, aptat et applicat . . . ita pariter in generatione animalium"; DG 162, W. 334. "Facultas enim pulli formatrix, materiam potius sibi acquirit, et parat; quam paratam invenit; videturque pullus haud ab alio feri, vel augeri, quam a se ipso"; 163, W. 336.

143. "Horum fabrica a parte aliqua, tanquam ab origine incipit; eiusque ope reliqua membra adsciscuntur; . . . haec per *epigenesin* fieri dicimus" (DG 161, W. 334); "Fit . . . ex puncto saliente, seu fundamento, corporis aedificium" (165f., W. 338).

144. ". . . videtur sanguis, tum generatione, tum natura et essentia, corde prius existere. Hoc enim ille utitur, seu instrumento. . . ." (DG 237, W. 411).

145. "Maius enim, et divinius inest in generatione animalium mysterium, quam simplex congregatio, alteratio, et totius ex partibus compositio; quippe totum, suis partibus prius constitutur, et decernitur; mixtum prius, quam elementa" (DG 167, W. 340).

146. "(Humido primigenio) videtur Natura . . . concessisse, quod materia primae . . . vulgo tribuitur; ut potentia nempe sit omnium formarum capax, actu autem formam nullam habet . . . Atque huic potissimum argumento innituntur, qui intellectum possibilem statuunt incorporeum; quia nempe susceptivus est omnium formarum sine materia"; DG 339, W. 513.

Cf. Aristotle, *De An.* III,4, 429a15–22: "[Mind] must therefore be without attributes, yet receptive of form, and *in potency be like objects but not the objects themselves*. And as the sense organ is related to the sensible, so must mind be to the thinkable. Therefore since mind thinks everything, *it must itself be unmixed*. . . . Thus it can have no other nature than this: *to be potential.*" (author's ital.) Averroes and Thomas developed the Aristotelian conception further to the doctrine of the "possible intellect." On the Galenic-Arab notions of the various elementary fluids, to which Harvey also refers (*ros primigenius, cambium, gluten*), see Pagel (1967), 257ff.

147. ". . . ex quo (procedentis generationis lege) mutato . . . ex inorganico, fit organicum . . . Non quidem transpositione aliqua, aut motu locali . . . sed potius disgregatione homogeneorum, quam heterogeneorum compositione"; DG 343, W. 517.

148. ". . . corpora similaria . . . elementa sua tempore priora non habeant, sed illa potius elementis suis prius existant (nempe *Empedoclis* atque *Aristotelis* igne, aere, aqua, et terra; vel *Chymicorum* sale, sulphure, et *Mercurio*; aut *Democriti* atomis) utpote natura quoque ipsis perfectiora"; DG 343, W. 517.

149. "Elementa . . . sunt . . . posteria potius; et reliquiae magis, quam principia. Neque *Aristoteles* ipsemet, neque alius . . . unquam demonstravit, elementa in rerum natura separatim existere . . ."; 343, W. 517. "Est tamen argumentum minus validum, corpora . . . naturalis ex iis generari vel componi primo, in quae corrumpuntur, vel solvuntur ultimo . . ."; 343f., W. 528. Cf. the *Praelectiones*: "The body as a whole is . . . called living . . . since it contains no part that is not living" (*corpus itaque ut totum dicitur . . . vivens . . . secundum quod tale nulla pars non vivens*; p.6).

150. Letter to J.D. Horst, 1 Feb. 1654; W. 613.

151. "Vita igitur in sanguine consistit, (uti etiam in *sacris nostris* legimus) quippe in ipso vita atque anima primum elucet, ultimoque deficit"; DG 202, W. 376. (In the marginal note, Harvey refers to Leviticus XVII, 11, 14.)

152. "corporis, animaeque commune vinculum est", DG 207, W. 381; "sanguinem esse partem genitalem, fontem vitae . . . sedemque animae primariam", 202, W. 377. In this context, Harvey refers explicitly to pre-Socratic doctrines of the animate nature of the blood (in Diogenes of Apollonia, Critias, and others) and criticizes the rejection of their view by Aristotle (201, W. 382). See also, Rüsche pp. 183, 249.

153. "si inciderit pavidus in lipothymiam, statim sistitur sanguinis effluxus. . . . Quid enim magis admirari, quam quomodo in omni affectu, appetitu, spe vel timore, corpus nostrum diversimodo patitur, et vultus evariatur ipse, et sanguis hic aut illuc subterfugiens videtur? . . . in verecundia, rubore . . . profunduntur genae: timore, infamia et pudore pallida facies, rubent auriculae . . . cupidine tactis adolescentibus, quam celeriter impletur nervus sanguine et erigitur . . ."; ER 152, W. 129.

154. ". . . omne namque animi pathema . . . ad cor usque pertingit, et ibi mutationem . . . in temperie et pulsu et reliquis facit"; DMC 84, W. 70.

155. ". . . non insunt vivi sanguini, sed a morte solum corrupto, et

iam dissoluto"; DG 213, W. 387. In this connection, it is significant for Harvey that it is a question only of *visible* components (*de . . . eius partibus ad sensum apparentibus*, 216, W. 390).

156. On this see also, Aristotle, *Meteorologica* IV 12, 389b27-390a11: "[All the homogeneous bodies] consist of the elements described, as their matter, but their essential nature is determined by their definition. This fact is always clearer in the case of the later products and in general all those things that are instruments, as it were, and have an end . . . . [T]he hand of a dead man . . . will be called a hand in name only, just as stone flutes might still be called flutes . . . All things are defined by their function; for each thing is what it really is when it is able to perform its function; an eye, for instance, when it can see."

157. "Non est opus . . . Deos in scenam advocare . . . : domi scil. nascitur, quod vulgo ab astris petimus"; DG 328, W. 502 (cf. ER 136, W. 116). "Sanguis . . . supra vires elementorum agit, cum iam pars primogenita et calor innatur existens . . . reliquas totius corporis partes ordine fabricat; idq. summa cum providentia et intellectu, in finem certum agens, quasi ratiocinio quodam uteretur"; DG 333, W. 507 [items in square brackets in translation added to Willis's version].

158. "Ad hunc etiam modum corpora omnia naturalia duplici rationi consideranda veniunt; nempe, vel prout illa privatim aestimantur, et intra propriae naturae ambitum comprehenduntur; vel prout instrumenta sunt agentis cuiusdam nobilioris, et superioris potestatis. . . . Quatenus vero praestantioris instrumenta sunt, et ab illo regulantur, . . . corpus aliud et divinius participare videntur, viresque elementorum excellere"; DG 334, W. 507f. "Nullum agens *primarium* . . . supra vires suas operatur: omne vero agens *instrumentale*, in agendo proprias vires superat"; DG 335, W. 509.

159. "In venis autem existens, quatenus est pars corporis, eademque animata et genitalis, atque immediatum animae instrumentum . . . Quatenus spiritus, est focus, Vesta, lar familiaris, . . . Sol microcosmi . . . not aliter certe, quam superiora astra . . . , servatis perpetuo circuitibus, inferiora ista vivificant"; DG 336, W. 510.

160. "Eodem ergo res redit, si quis dicat, Animam et sanguinem, aut sanguinem cum anima, vel animam cum sanguine, omnia in animali perficere"; DG 227, W. 511.

161. "istoque circuitu, illorum genus aeternitatem . . . consequatur"; DG 56, W. 225. "Facit . . . hic circuitus . . . genus sempiternum; . . . continuata perpetuo serie, ex individuis caducis et pereuntibus, immortalem speciem producunt"; 115, W. 285. See also, Aristotle, II,1, 731b18–51.

162. Brunn calls attention to this correspondence (49); cf. DMC 34, W. 29 (*sperma animalium omnium . . . et spiritus palpitando exit, velut animal quoddam*—thus like the *punctum saliens*); DML p. 94; also DG 139, W. 311 (*[semen] copioso spiritu impulsore opus habet*—namely, in ejaculation).

163. "Eademque inquisitio est, quid sit in ovo, quod illum foecundum reddit, . . . quid in . . . ovario; quid in foemina; quid denique in semine, et gallo ipso: et, Quid sit in sanguine, punctoque saliente . . . , unde postea reliquarum partium ortus, fabrica, et ordo promanet; Quid in pullo ipso,

unde robustior, et agilior fiat, citius adolescat", etc.; DG 187, W. 361. This principle already played the decisive part in cardiac activity: it occurs *"ab interno principio, regulante natura."*

In these sections, Harvey does not refer explicitly to the formal cause; in a sense, it seems to be contained in the first efficient cause. The *De Conceptione,* an appendix to the *De Generatione,* confirms this: *"Inest enim . . . in omni efficiente, ratio finis"* (401, W. 583); but this *"ratio"* is also described as *"forma sine materia"* of the living thing. To be sure, Aristotle, too, often speaks in the context of conception, especially of the semen as efficient cause (although it must also transfer the form; DGA II,1); nevertheless, in Harvey's interpretation, the emphasis on the active *creative principle* appears plainly over against the ancient principle of order and form.

164. *"Prima* igitur efficientis proprie dicti . . . conditio . . . est, ut sit foecundans primum et principale, unde media omnia foecunditatem . . . accipiant"; DG 188, W. 362f.; ". . . ut instrumenta omnibus intermediis successive aut motum impertiat, aut aliter iis utatur, ipsum vero nulli inserviat"; 190, W. 364. Cf. Aristotle, *Phys.* VIII,5, 258b8f.: ". . . the primary mover is unmoved. . . ."

165. ". . . efficientis primi (cui ratio futurae prolis inest) eam legem esse, ut, cum proles mista appareat, et ipsum quoque mistae naturae sit"; DG 190, W. 364; ". . . (ut) effecto, opereve suo tanto excellentior sit, quanto architectus operi suo praestat"; 192, W. 267.

166. This is, in fact, a regular theme of the Aristotelian doctrine of nature; it points to the primary reality of form that underlies every natural process.

167. ". . . ut primum efficiens in pulli fabrica artificio utatur, et providentia, sapientia item, bonitate, et intellectu, rationis animae nostris capitum longe superantibus"; DG 191, W. 366.

168. ". . . solem accedentem, et recedentem, sequuntur ver et autumnus; quibus plerunque temporibus, animalium generatio, et corruptio contingunt; 192, W. 367; also 193f., W. 368; 196, W. 370. (Incidentally, numerous passages indicate that Harvey ignored the new astronomy and proceeded from a moving sun (DMC 56, W. 46; DG 57, W. 226; DG 115, W. 285f) and an earth at rest (DML 54, W. 56; DG 44, W. 213).

In the section under discussion here, Harvey cites a passage from Aristotle, *De Gen. et Corr.* (II, 10, 336a32ff.): "The cause of generation and corruption is . . . the motion along the inclined circule; for in this motion there is continuity and duality of movement. For if coming-to-be and passing-away are always to be continuous, there must be some body always being moved (in order that changes may not fail) and move with duality of movement (in order that both changes, not one only may result) . . . For of contrary effects there are contrary causes." (. . . *ortus ad interitus caussa . . . est . . . obliqui circuli latio; ea namque et continua est, et duobus motibus fit; nam si generatio, corruptiove futura sit semper continua, semper quippiam quidem moveri necesse est; . . . sed duobus, ut non altera duntaxat eveniat. . . . Contrariorum enim caussae sunt contrariae*; DG 192, W. 367).

As early as the concluding section of the *De Motu Locali,* Harvey brings rhythm and *vicissitudo* into relation with systole and diastole and with the change of seasons: ". . . and so relaxation is a part of action, like the systole and diastole of the heart. Hence frogs and swallows hide in winter and show no movement" (. . . *et sic relaxatio pars actionis est ut cordis systole diastole. Unde frogges, swallowes hiemi latent et motum no edunt*; p. 152). Contraction and relaxation, activity and passivity, but also generation and corruption are equally components of the rhythm of the living.

169. "Man begets man, and so does the sun"; *Phys.* II,2, 194b13f. ". . . the cause of man is not only the elements, fire and earth, as matter, and the peculiar form, but also some other external cause, e.g. the father, and besides that the sun and the inclined circle, and the last, indeed, not as matter or form . . . but as moving cause"; Aristotle, *Met.* XII,5, 1071a13ff. Thus, the subordination of the father as instrument is not Aristotelian, but is Harvey's reinterpretation.

170. ". . . necesse est fateamur, in generatione hominuis, causam efficientem ipso homine superiorem et praestantiorem dari: vel facultatem vegetativam, sive eam animae partem, quae hominem fabricat, et conservat, multo excellentiorem, et diviniorem esse, magisque similitudinem Dei referre, quam partem eius rationalem . . ."; DG 194, W. 368. Surprisingly, Willis's translation gives the opposite sense: "it is imperative even in the generation of man to admit an efficient cause, superior to, and more excellent than man himself: *otherwise* the vegetative faculty . . . *would have to be accounted far more excellent and divine* . . . than the rational part of the soul . . ." (author's italics). The *vel* (in medieval Latin often used to give emphasis: "in other words," "or indeed") is certainly here not to be understood in the sense of "*aliter*" (otherwise). Moreover, the parallel passage at p. 228, W. 402 speaks against Willis's reading: "*vegetativae* operationes potius videntur arte . . . et providentia institui; quam animae rationalis, mentisve actiones." I therefore accept E. Lesky's interpretation (p. 373).

171. ". . . Natura . . . et anima vegetativa . . . movent, nulla facultate acquisita, (sicut nos) quam vel artis, vel prudentiae nomine indigitemus; sed tanquam fato, seu mandato quodam secundum leges operante; simili nempe impetu, modoque, quo levia, sursum, gravia, deorsum feruntur . . . non autem . . . disciplina, consilio quicquam agunt"; DG 195, W. 369.

172. "Jovis omnia plena," DG 227, W. 401; "animae plena sunt omnia," 113, W. 284.

173. We have to count, here in particular, the apparently immaterial, "contagion-like" action at a distance of the male seed on the already living, but as yet unfertilized, egg; cf. DG chaps. 26–41. On this whole problematic, which I cannot explain at greater length here, I refer the reader to E. Lesky, E. Gasking, D. Goltz, "The Empty Uterus: On the Influence of Harvey's *De Generatione Animalium* on Theories of Conception," *Med.Hist.J.* 21 (1986), pp. 242–68; F.J. Cole, *Early Theories of Sexual Generation*, Oxford, 1930, pp. 135ff.

174. Thus, as E. Lesky writes, the vegetative soul has "gone far beyond the Aristotelian force of the same name, with which it no longer has anything but the name in common" (Lesky, p. 374, translated by the translator).

175. Harvey presents this polarity in many passages: "... mas et foemina separatim non sunt generativi, sed uniti ... ab utroque, tanquam uno ... (ovum) educitur," 126, W. 297; "uterque, tam gallus, quam gallina pulli parens dicendus est; sunt enim ambo principia ovi necessaria," 136, W. 307f; "quod in ovo pullum efficit, mistae naturae, (tanquam aliquid ex ambobus vel unitum, vel compositum) ... est"; 189, W. 363; also 56, W. 225; 116, W. 286. On the circle of hen and egg, cf. note C161 or p. 101, W. 271; "nam haud facile dixeris, utrum ovum pulli ex eo nati gratia; an hic illius causa facta fuerit." Here too, Harvey speaks of a *vicissitudo*: "ita ... caducae res mortalium, *alternis individuorum vicissitudinibus, ... perennant*"; 57, W. 226.

The *higher development of heart and blood through polarity* has already been discussed. On reproduction see p. 137, W. 309: "... in the generation of the most perfect animals, where *principles are distinguished*, and the seminal elements of animated beings are divided, a new creation is not effected save by the concurrence of male and female, or from two necessary instruments" ... *natura, ad perfectiora et praestantiora animalia procreanda, pluribus instrumentis ... necessario utitur ... in perfectissimorum animalium generatione,* unde principia istaec distinguuntur, *et haec animalium semina divisa sunt; non nisi ex mare et foemina, tanquam duobus instrumentis requisitis, procreantur;* (author's italics).

176. DG 334, W. 507f.; for Latin text, see note C-158.

177. DG 150, W. 322: "Plura enim et efficaciora fabricando animali, eidemque conservando requiruntur, quam ad eius perniciem. ... Cernitur in generatione rerum ... omnipotens Deus ... ; mille autem modis mortalia cuncta ad perniciem *sua sponte* ruunt" (author's italics).

178. "Quinetiam foemina potiori iure efficiens videatur: ... inter animalia, foeminae nonnullae sine mare procreant"; DG 118f., W. 289. "Generat scil. foemina, sed aliquatenus; et maris coitus requiritur, ut facultas istaec generandi in foemina perficiatur"; 131, W. 302, among other places. On this, see H. Fischer, "Die Geschichte der Zeugungs- und Entwicklungstheorie im 17. Jahrhundert," *Gesnerus 2* (1945): 49–80, p. 58.

179. "Isti duo motus, auricularum unus, alter ventriculorum, ita per consecutione fiunt, ... ut ambo simul fiant, *unicus tantum motus* appareat, *praesertim in calidioribus animalibus*"; DMC 37, W. 30 (author's italics). Cf. DG 126, W. 297: "... ab utroque, tanquam uno ... (ovum) educitur"—and this polarity too characterizes the "more perfect" warmer animals.

180. For the first view, G. Whitteridge may be cited as typical (pp. 226ff., esp. p. 232), for the second, C. Webster (p. 272f.).

181. Pagel (1967), pp. 261 ff., 336; (1976) p. 52.

D

THE MECHANICAL ASPECT OF THE CIRCULATION: DESCARTES AND HIS FOLLOWERS

I. CIRCULATION AND PHYSIOLOGY IN DESCARTES

<u>Summary</u>. *(a) In contrast to Harvey's conception of science, Descartes does not determine the first principles of his natural philosophy in inductive fashion, but through an immediate conceptual assertion: his basic concepts are geometric matter devoid of qualities and the motion of its elementary particles. The concrete phenomena count as real only insofar as they can be derived from these abstract concepts. Deducibility is also the presupposition of practical production or at least of manipulation. Descartes sees in this the true goal of science. As the most general forms of natural as well as of artificial processes, the laws of nature become the most important principles in the study of nature. They constitute a world of purely mechanical causal connections to the exclusion of magical or teleological explanations.*

(b) For physiology, this means the dualistic separation of bodily sensation and psychological experience from physical events in the body. The machine model of the organism, that is, the replacement of Harvey's autonomy of the organs by mechanistic automatism and the substitution of externally guided reflexes for self-movement, is the basis for the first thoroughly "soulless" physiology, which is thus "scientific" in the modern sense.

(c) In what follows, we examine how, primarily through conceptual reinterpretation, Descartes isolates the material level of Galenic physiology and thus, without gaining essentially new knowledge, nevertheless interprets bodily processes in a wholly new way. Thus in Descartes the calor innatus, the "vital heat," becomes an exothermic reaction of particles, that is, an event indistinguishable

from purely inorganic, physico-chemical processes. This reaction provides the driving force for the cardiac motor and thus for the whole bodily machine. On the other hand, the reinterpretation of the spirits as a neuronal stream of particles makes possible a mechanical-cybernetic model of the control and movement of the organism.

But it is the circulation that produces the decisive connection between triggering and control, now conceived as a "transmission belt", but also as a circular feedback system. Instead of the vital independence of the organs, as in Harvey's conception, we have in the end their total subordination to the central nervous system; and in the Cartesian theory of the passions, the heart too—in the first instance only, in its role in the origin of the emotions—is tied to central control. This theory explains kinesthetic feelings as physical events determined by the nerves and the humors whose psychological equivalents are simply projected onto the region of the heart. Insofar as he robs the heart of its central significance for human self-consciousness, Descartes provides the necessary condition for the purely scientific study of the heart after his time.

René Descartes was among the first to support Harvey's theory of the circulation. He accepted it virtually sight unseen and integrated it into his system of natural philosophy though with an account of cardiac action contrary to Harvey's.[1] Later, through his physiological expositions in the *Discourse on Method* (1637), the *Principles of Philosophy* (1644), and the *Passions of the Soul* (1649), as well as through his extensive contacts and correspondence, he became probably the most influential proponent of the new doctrine—at a time when most medical authorities still held fast to the traditional physiology and Harvey found himself faced with considerable opposition. How far Descartes's influence affected not only the acceptance, but also the interpretation of the theory of the circulation will be investigated in Chapter II. First, it is necessary to show how, on the basis of the circulation, Descartes develops an entirely new point of view in physiology, which differs from Harvey's on essential points. It is impossible even to think of such a project, however, except within the framework of the Cartesian system of natural philosophy. For this reason, as I did with Harvey, I shall first sketch in its main features Descartes's conception of science and his vision of nature.

a. CARTESIAN SCIENCE

The science of nature must rest on first principles; in this, Descartes agrees with Harvey. However, he does not arrive at these principles inductively by retracing and analyzing what is given in the first instance in sense perception, but directly through theoretical reflection: "true" is what is known "clearly and distinctly" to the intellect.[2] In a famous passage in the *Meditations*, Descartes observes how wax that is melted changes in all its *perceptible* properties and infers from this that

> . . . even bodies are not strictly perceived by the senses or the faculty of imagination but by the intellect alone, and that this perception derives not from their being touched or seen but from their being understood. . . .[3]

However unreliable the sense perception of corporeal things, at least ". . . they possess all the properties which I clearly and distinctly understand, that is, all those which, viewed in general terms, *are comprised under the subject-matter of pure mathematics*" (AT VII 80, CSMK II, 55; author's italics).

In this way, Descartes arrives at first principles of natural philosophy fundamentally different from those accepted by Aristotle or Harvey. What we have is, on the one hand, the purely geometrical *extension* of matter, devoid of qualities, and on the other, the local *motion* of particles that we conceive of, but cannot see.

> For I freely acknowledge that I recognize no matter in corporeal things apart from that which geometers call quantity, and take as the objects of their demonstrations, i.e. that to which every kind of division, shape and motion is applicable. Moreover, my consideration of such matter involves absolutely nothing else apart from these divisions, shapes and motions; and even with regard to these things, I will admit as true only what has been deduced from indubitable common notions so evidently that it is fit to be considered as a mathematical demonstration. And since all natural phenomena can be explained in this way, . . . I do not think that any other principles are either admissible or desirable in physics (PP II, 64, AT VIII 1 78–79, CSMK I 247).

Thus the principles are not, as in Harvey's conception of science, inherent in the phenomena, i.e. potentially contained in sense perception; on the contrary, they must be grasped in thought precisely through the *avoidance* of sensory experience. For the ". . . first and main cause of all our errors" (PP I, 71, AT VIII 1, p. 35; CSMK I 218) lies in the naive identification of sense perception and truth. In the *Dioptrics* (1637), Descartes explains this in more detail.[4] The soul does not receive in the brain the pictures and forms of the things themselves; for sensations of the most divergent kinds all arise through the same mode of transmission, namely motions (of particles) in the nerves. It is *signals* or *symbols* that transmit to the soul certain information important for the maintenance of the body without thereby saying anything about the true nature of things. The soul, in its brain, is dependent on the nerves, *like a blind man on his stick*, which communicates to him different properties of objects through his movements alone. So, it comes about that

> . . . we think we see the torch itself, and hear the bell, and not that we have sensory perception merely of movements coming from these objects.[5]

Thus what is given in sense perception cannot be the touchstone of the general principles of science as it is for Harvey. It is exactly the opposite. If the perceptions of sense, kinesthetic sensations, or emotions contradict the principles, then we have every reason to call the *former* into question.[6] Descartes's critical attitude to experience is not untypical for modern science; similar remarks can be found in Bacon, Galileo, Kepler, Newton, and others. Despite his mistrust of the senses, however, Descartes cannot be called a "deductivist" in the modern sense. Not only does he use the word "deductio" sometimes for an explanation, even for an inductive argument; he also accepts any inference, whether deductive or inductive, on condition that it promises the best available access to the truth. As early as the *Regulae ad directionem ingenii*, Descartes sets the intuitus, or natural light of reason, above any syllogistic inference.[7] Fundamental doubt as the starting point of Cartesian science possesses, as not the least of its purposes, the function of eliminating from nature all qualities perceptible through the senses in order to reduce such qualities to the configuration and motion of submacroscopic particles. These particles, which were considered by Harvey to be mere *entia rationis*, now become, *precisely as such*, the real constituents of all phenomena in nature and in man.[8]

Nevertheless, despite its deductive procedure, science repeatedly encounters branchings of the causal sequences that occur in the ongoing course of natural processes. Here, the decision, which is the "correct" one among many possible worlds, demands supplementary, empirical points of reference. Hence, Descartes presents his cosmological and physiological doctrines in the first instance only as "hypotheses"; he draws up "models" in the modern sense, models of the universe as well as of the body.⁹ He does this, on the one hand, because of the limitations of the knowledge so far available—the new science is also a *research program* (*Disc.* 6)—on the other, as a more conciliatory way of introducing a new thought style, not least with an eye to religious orthodoxy. Basically, however, the question of the real context is nevertheless secondary compared to the hypothetical reconstruction of reality; for the latter:

> . . . yields just as much practical benefit for our lives as we would have derived from knowledge of the actual truth {because we shall be able to use it just as effectively *to manipulate natural causes so as to produce the effects we desire.*} (PP III 44, AT VIII-1 99; bracketed clause in French only, AT IX-2 123, CSMK I 255; author's italics.)

The model may well remain a model; its agreement *on principle* with reality is sufficient for the reconstruction in question. For the task of science in Descartes's case lies, not in the pure knowledge of reality, in the way that Harvey conceived it as the goal of science, but in the process of carrying out deductions as the actual production of real results. In other words, the task of science consists in the technical molding and subjugation of nature:

> I shall think I have achieved enough provided only that what I have written is such as to correspond accurately with all the phenomena of nature. This will indeed be sufficient for application in ordinary life, since medicine and mechanics, and all the other arts, which can be fully developed with the help of physics, are directed only towards items that can be perceived by the senses and are therefore to be counted among the phenomena of nature (PP IV 204, AT VIII-1 327, CSMK I 289; cf. *Disc.* 6).¹⁰

Thus, Cartesian science is concerned with the comprehension not of things in themselves, but of the *forms of their possible transformation*. Hence, the most general of these forms, the *laws of nature*, belong among these most important principles.

These "laws" no longer refer, as with Harvey, to cyclically recurring regularities which characterize the purposive behavior and creativity of nature; instead, they are absolute laws of the movement of bodies, ordained at the beginning by God to reign over the material world. The concept of "law" in this modern form appears for the first time in Descartes, but soon acquires general currency.[11]

The Cartesian laws of nature decree that every body must persist in its condition of rest or motion (first law, AT VIII-1 62); this motion is uniform and rectilinear (second law, AT VIII-1 63); only the mechanical effect of pressure and impact of other bodies leads—again in a law-like manner—to changes of this condition (third law, AT VIII-1 40). In this way, the "natural motion" of Aristotelian physics, i.e. the specific type of self-movement characteristic of different kinds of material, is set aside. As a matter of principle, *movements are no longer end-directed, but forced and passive*; this holds also for the realm of the living, so that it would now be impossible to speak, for example, of a "centripetal tendency" or of a "spontaneous self-movement" of the blood. For a body is either already in motion, or begins to move through the impact of another body, which transfers to it a part of its movement. *But the total quantity of motion in the universe remains constant*, as expression of the constancy of the lawgiver, God himself (PP II 36, AT VIII-1 61-2, CSMK I 240)

A world of constantly moved particulate matter and its laws: these are therefore the principles from which, logically as well as in their historical origin, all the phenomena of nature can be derived—inanimate as well as animate, and in that order:

> For God has established these laws in such a marvelous way, that even if we supposed he creates nothing beyond what I have mentioned, and sets up no order or proportion with it, but composes from it a chaos ... the laws of nature are sufficient to cause the parts of this chaos to disentangle themselves and arrange themselves in such good order that they will have the form of a quite perfect world.[12]

Hence, in the Cartesian system *cosmogony*, or as its counterpart for the present state of the world, *astronomy*, becomes the paradigmatic science. Ever since antiquity, it had been considered the model-building science *par excellence*, which had to concern itself not with the comprehension of its objects in themselves, but with "saving the phenomena." For that reason, as we have seen, Harvey regarded it with skepticism. But it was precisely in this feature of astronomy, and in the pure lawfulness that rules the heavens, that Descartes found the key to a universal science:

> ... The discovery of this order is the key and foundation of the highest and most perfect science of material things that man can ever attain. For if we possessed it we could discover *a priori* all the different forms and essences of terrestrial bodies, whereas without it we would have to content ourselves with guessing them *a posteriori*.[13]

This universal science can be realized only as mathematical physics. Such a science no longer recognizes any basic difference between living and dead objects; life becomes a "variant of the possibilities of the lifeless."[14] To achieve this, it is necessary above all to deny every form of natural causality that would transcend the laws of nature, or "transubstantiate" matter—magical or analogical relations, qualitative changes of form (PP IV 187, AT VIII-1 314-315, CSMK I 278-9 [given in part]), but especially *final causality*. The goal of a universal science is achieved if there is no longer any goal-directedness within its sphere of validity. In this context, the greatest obstacle would be an inner finality given to the world by the creator himself in its very creation—a purely immanent, "unconscious" teleology in the Aristotelian sense Descartes considers quite impossible. Thus, the ends of God's creation are as impenetrable as his efficacy in the creation of mobile matter and in the proclamation of its laws is calculable:

> When dealing with natural things we will, then, never derive any explanations from the purposes, which God or nature may have had in view when creating them {and we shall entirely banish from our philosophy the search for final causes}. For we should not be so arrogant as to suppose that we can share in God's plans. We should, instead, consider him as the efficient cause of all things ... (PP I 28, AT VIII-1-15, CSMK I 202, bracketed clause in French only, AT IX-2 37).

Granted, God is the *primum efficiens* of Nature, as he was with Harvey, not now, however, in the sense of a *vis enthea* present in all things, an engendering, planning and shaping principle, an "unconscious providence," but only as the cause of a world of closed physical causality—a cause itself no longer physical—or as the divine geometer of a law-like, calculable nature.[15]

To extend these physical laws to the sphere of the living—that is the task Descartes sets himself in his physiology. But the first vehicle for this reinterpretation of organic nature is the circulation of the blood, which, as we shall see, no longer imitates the eternal, cyclical self-movement of the heavenly bodies, but the mechanical vortex-system of the Cartesian cosmos.

b. THE BODY WITHOUT SOUL

Like all living beings, the human body, too, is now integrated into a fundamentally uniform physical world that is filled everywhere with the same matter (PP II 22f., AT VIII-1 52-3, CSMK I 232-3). No special lawfulness, no relation of parts to whole, delimits the space of the living body from the general space of *res extensa*. Thus, the Cartesian dichotomy of body and mind cuts through the human being—two mutually exclusive substances without a connecting ground: "my I, that is my mind—for that is all I now take myself to be" (Med. III, AT VII 51, CSMK II 35) and opposed to it "the whole mechanical structure of limbs ... , which can be seen in a corpse, and which I called the body" (Med. II, AT VII 26, CSMK II 17). The soul loses its presence in the whole organism and is reduced to the *anima rationalis* in the brain.

> ... [W]hen we try to get to know our nature more distinctly we can see that our soul, in so far as it is a substance which is distinct from the body is known to us merely through the fact that it thinks, that is to say, understands, wills, imagines, remembers and has sensory perceptions, for *all these functions are kinds of thought*. The other functions which some attribute to the soul, *such as moving the heart and the arteries*, digesting food in the stomach and so on, do not involve any thought, and are simply bodily movements (DCH AT XI 224-5, CSMK I 314-5, author's italics).[16]

But even voluntarily executed bodily movements can be distinguished from those that are involuntary only by a *preceding* mental cause—and not by their physical mechanism, their "common motoneuronic pathway." On the other hand, in bodily processes, the causal chain running through the particles can continue as far as the brain and can trigger feelings of sense or emotion in consciousness. It is neither collaboration nor hierarchical organization, but *reciprocal external influence* that characterizes the relation of the two substances, mind and body.

This process of crossing the threshold of consciousness in

either direction must be sharply separated from the physical movements in the body itself, which depend only on the order and configuration of its parts. For the fact that they remain in the unconscious means only that they occur "on their own," without a vegetative, vital, or moving function of the soul:

> ... [W]hen all the bodily organs are appropriately disposed for some movement, the body has no need of the soul in order to produce that movement... (DCH AT XI 225, CSMK I, 315).

Thus in the explanation of nature, Descartes, like Harvey, uses the principle of parsimony: entities are not to be multiplied beyond necessity. But while Harvey is trying by this means to reduce the complicated soul-body hierarchy of Galen to the level of the vital, Descartes takes another path. He uses precisely the external opposition of the various intermediates already present in Galenism in order to separate the organism as such from the soul, to posit the lowest, material level in it absolutely, and to explain vegetative-vital processes exclusively in terms of that lowest level. What was needed for this purpose, as we shall see, was chiefly no more than a reinterpretation of the "psycho-material" explanations that were already available. The result is not Harvey's *autonomy of the vital*, but *mechanical automatism*.

> So I will now try to ... give such a full account of the entire bodily machine that we will have no more reason to think that it is our soul which produces in it the movements which we know by experience are not controlled by our will than we have reason to think that there is a soul in a clock which makes it tell the time (DCH, AT XI 226, CSMK I 315).

The mechanical no longer represents an element within vital organization; through the *paradigm of the automaton*, it embraces the organism itself. In this context, the analogy of the clock illustrates the continual running of bodily functions, especially their triggering mechanism in the heart. In other words,

> ... the difference between the body of a living man and that of a dead man is just like the difference between, on the one hand, a watch or other automaton (that is, a self-moving machine) when it is wound up and contains in itself the corporeal

> principle of the movements for which it is designed, ... and the same watch or machine when it is broken and the principle of its movement ceases to be active (PA I 6, AT XI 331, CSMK I 329-330).

Other comparisons offer models for the *regulation* and the *reactions* of the body: the hydraulic automata of the gardens at Versailles, which can be regulated by a distributor (TH AT XI 130, CSMK I 100) or the organs over whose keyboards the inner stream of air can be guided and sounds released (TH AT XI 165).

Two important reinterpretations lie hidden in these comparisons. Descartes calls the automaton "a machine that moves itself"; however, an automaton does not *set itself in motion* on its own. Thus, the comparison with living movement suggests that, in the last analysis, living movement, too, is effected *externally*, in other words, that it is brought about *by compulsion* through initial "winding up" (as with the clock) or through repeated stimuli (as in the case of the organ). Self-motion in the spontaneous or active sense is thus excluded, and what Harvey sees as an "answer," as reaction of the sensitive tissue, becomes, through mechanical association, a physical occurrence that follows necessarily from the laws of nature: movement is then nothing but *being moved*.[17]

The other reinterpretation concerns the inner purposiveness of the organism. Like the teleology of a machine, it is attributed to a plan provided in advance by an external source:

> God made our body like a machine, and he wanted it to function like a universal instrument, which would always operate in the same manner according to his laws (AT V 163, CSMK III 346. with modification).[18]

Granted, the program of renouncing teleology cannot yet be carried through completely in the sphere of the living. However, insofar as teleology is transferred to the planning of the divine architect, the decisive step has already been taken. The body is no longer in any way "its own goal," but only a means or instrument whose goal remains entirely external.[19] Moreover, Descartes permits no doubt that this solution is a temporary one, and he does suggest an attempt at a mechanistic embryology:

> Nevertheless, if we want to understand the nature of plants or of men, it is much better to consider how they can gradually grow from seeds than to

> consider how they were created by God from the beginning of the world. Thus we may be able to think up certain very simple and easily known principles, which can serve, as it were, as the seeds from which ... everything we observe in this visible world could have sprung (PP III 45, AT VIII-1 100, CSMK I 256).

Even though at this time organic evolution does not yet lie in the field of vision, still, we already meet here, in principle, the idea behind the modern theory of evolution, that is, the idea of attributing the apparent purposiveness of nature to external, physical boundary conditions.

Now, since purposiveness is imposed on the organism from without, such purposiveness is now distinguished from the functionality of technical automata, not on principle, but only through its greater complexity (PP IV, 203, AT VIII-1 326, CSMK I 288); *the distinction between "natural" and "artificial" becomes untenable.*

> And it is certain that all the rules of Mechanics also belong to Physics ... , so that all things that are artificial, are thereby also natural. For, to take an example, when a clock marks the hours by means of the wheels of which it is made, that is no less natural than it is for a tree ... to produce its fruits (PP IV 203, French text, AT IX-2 321-2; cf. CSMK I 288).[20]

Descartes, too, recognizes the extraordinary workmanship of many animals, which in Harvey's eyes allow the unconscious operation of nature to appear so superior to human reasoning; but for Descartes, the consequence is radically different:

> ... what they do better [than we do] does not prove that they have any intelligence. ... It is nature that acts in them according to the disposition of their organs. In the same way a clock, consisting only of wheels and springs, can count the hours and measure time more accurately than we can with all our wisdom (Disc. 5, AT VI 58-9, CSMK I 141).

Harvey's "unconscious providence of nature", or the *ingenium connatum* of animals, is for Descartes only a complex program of movements. Let us now turn to those programs in some detail.[21]

C. THE PHYSIOLOGICAL MECHANISMS

1. Motion of Blood and Heart.[22] The chain of physiological movements of particles begins with digestion in the stomach, a process both of fermentative disintegration and of the filtering of nutriment through "pores" in the wall of the intestine (TH, AT XI 121). In this context, it is crucial that, on Descartes's view, such "chemical" processes as the decomposition of nutriment differ in no way from "physical" processes since he replaces qualitative changes of form by corpuscular movements. Digestion is explained on the model of slaking lime, the effect of nitric acid on metals, and the fermentation of damp hay (loc.cit.), processes that, in the Principles, Descartes compares to the origin of fire. It is a matter of basically similar, namely, heat-producing, "exothermic" reactions.

> That indeed the particles of some spirits or fluids can produce fire when they enter into the passages of a hard or even of a fluid body is shown by moist hay when it is shut in, or by lime when water is poured over it, or by fermentations and many fluids known to chemists, which heat up when mixed with one another and even burst into flame (PP IV 92, AT VIII-1 256).

According to Descartes, there are three kinds of matter (PP III 52, AT VIII-1 105, CSMK I 258 [in part]). The first element, also called ether (PP IV 100, AT VIII-1 260), consists of the finest and most rapidly moving, invisible particles; it forms the substance of the stars. The second element consists of somewhat slower "spherical particles" whose movement is perceptible as light. Finally, to the third element belong the variegated particles of air, earth, and water that constitute the matter of earth (PP IV 33, AT VIII-1 220); they are constantly surrounded by the finer particles of the first and second kind. If, through friction, impulsion, etc., the second element is temporarily driven out of the interstices of terrestrial matter, then those second element particles are exposed to the unqualified motion of the first element and are carried away by it. A fire has arisen, whose heat consists solely in the increased motion of its parts, and which needs the constant addition of terrestrial particles for its conservation (PP IV 80, AT VIII-1 249-250).

However, if such a reaction takes place more slowly, then there is no accompanying emission of light since the "celestial spherical particles" of the environment are not set into sufficient motion to appear as light; this holds also for the above-mentioned processes of fermentation, the slaking of lime, and so on. Nevertheless, these are to be explained on the same model as visible fire. In the composition or mixture of two materials, the one that is moved more forcefully pushes out the particles of the second kind from the interstices or "passages" of the other with the result that the latter is decomposed by the first element (PP IV 90 ff., AT VIII-1 254 ff.). It is decisive here, on the one hand, that Descartes no longer considers fire a special element, but takes it to be special form of the heat-producing reaction of particles—in the sense of a general concept of "combustion" or of modern "oxidation." But on the other hand, it is equally important that in this way a physical solution can for the first time be given for the old problem of the relation of fire and vital heat; for "fermentation," too, which is now the basis of physiological processes, is no longer anything specifically organic or, as it was for contemporary alchemists, a qualitatively ennobling process, but arises merely from the inorganic world of particles.

Decomposed by such processes and filtered through the walls of the intestines, the nutriment then mixes with the blood of the portal vein and arrives at the liver, where it is transformed into blood through further filtering and chemical refinement (TH, AT XI 121f.). The blood is no special "life stuff"; on the contrary, it is "nothing but ... a collection of small particles of food" (DCH, AT XI 250), a mixture of materials. This mixture then becomes the fuel for the fire that is maintained in the heart (Disc. V, AT VI 48-9, CSMK I 135). The *calor innatus*, the vital heat, now consists only in a continuous material process of combustion. Here too, it is a matter of a "fire without light": the inorganic chemical reaction proceeds according to the familiar pattern in the form of an instantaneous fermentation.[23] " . . . . a kind of fire that is without light, like that started in aqua fortis when sufficient steel dust is added, and like that of all fermentations." (" . . . une espèce de feu, qui est sans lumière, semblable . . . celuy qui s'escite dans l'eau forte, lorsqu'on met dedans assez grande quantité, de poudre d'acier, et . . . celui de toutes les fermentations.")

This reaction is now the sole cause of cardiac activity, and therewith at the same time, "*the first principle and motor of our whole machine [i.e. of the body]*," (DCH AT XI 228; cf. Disc. 5, AT VI 46-7.) In these circumstances, the heart itself plays only a passive role: it is inflated by the blood that is heating and expanding, and this blood in turn opens the valves of the aorta and the pulmonary vessel and streams into the vessels. Thus, inappropriately, Descartes attributes the flow of blood from the heart to diastole, while in systole the heart is emptied and collapses (DCH AT XI 241-5, CSMK I 316-319; Disc. Pt. V, AT VI 49-50, CSMK I 135-136); the valves prevent the blood from backing up "in agreement with the laws of mechanics."²⁴

Nevertheless, some heated blood has remained in the pores between the cardiac fibers, and this now mixes with the blood freshly streaming in from the auricles and serves as *ferment* for the next reaction (DCH, AT XI 231, 281). Thus in the continuous accession of blood, there arises a ceaseless chain reaction like that of a combustion engine, and there is "no other heat in the heart but this movement of the blood particles" (DCH, AT XI 281). In this process, the passage of the blood through the lungs serves for the recondensation of the vaporized blood, which now, in a second stage of the reaction, can engender more heat in the left ventricle. Thus, Descartes reinterprets the Galenic *concoctio* as a repeated distillation, which also explains the difference between arterial and venous blood.²⁵ Cf. DCH, AT XI 236.

Since the mechanistic approach stands or falls with the physicalistic explanation of the primary movement of the body, Descartes defended the theory of cardiac action he had developed against all comers. He writes repeatedly that it is

> so important to know the true cause of the heart's movement that without such knowledge it is impossible to know anything which relates to the theory of medicine (DCH, AT XI 245, CSMK I 319).

In the *Discourse,* Descartes calls the action of the heart a mechanism that runs with the same necessity ". . . as the movement of a clock follows from the force, position and shape of its counterweights and wheels" (Disc. 5, AT VI 50, CSMK I 136 [modified]). In contrast, according to the presuppositions of *Harvey's* theory, " we must imagine some faculty which causes this movement . . . . ..[and] we shall also have to suppose that there are additional faculties which change the qualities of the blood while it is in the heart" (DCH, AT XI 243-4, CSMK I 318).

Descartes correctly sees in Harvey's conception of cardiac contraction a decisive difference from his own system; however, mistaking Harvey's intentions, he sees it as a relapse into the Galenic schema of explanation with the help of the soul. Harvey does indeed speak occasionally of a *vis pulsifica* (see note C 77); but, as we have seen, he understands contractility and sensibility as *specifically vital qualities of the tissues*—a point, of course, that Descartes would not have accepted either. There is agreement between them on the view that the blood undergoes a change in the heart, namely, the change from venous to arterial blood. But without a chemical theory of cardiac action, Descartes considers the recourse to a *virtue* or *vital faculty* inevitable, and Harvey does indeed speak of a "virtue" of the heart that renews the blood. The focal point of the Cartesian theory consists in the fact that the same material process makes the heart an organ of movement as well as of "metabolism," so that

> ... this expansion alone is sufficient to move the heart in the way I have described, and also the change the nature of the blood in the way which observation shows to be the case (DCH, AT XI 244, CSMK I 318).

In Harvey, too, the regenerating *virtus* of the heart is inseparable from its action, i.e. the violent agitation of the blood; this connection in turn rests on the specific capacity of the blood for "vital heat." Thus, both Harvey and Descartes bring together the kinetic and the qualitative aspect of cardiac action—but the one under the banner of vitalism and the other of a physico-chemical reinterpretation.[26]

Despite some erroneous explanations in its details, Descartes's fundamentally novel point of view proved, in its very novelty, persistently influential. This is especially clear in the correspondence with Vopiscus Plemp (1601-1671), Professor of Medicine in the Dutch University of Leuven, which was published in 1644.[27] In fact, Descartes was able to persuade Plemp to accept Harvey's doctrine of the circulation. Plemp still holds fast to the Galenic *pulsative faculty* of the heart and objects to Descartes that the continued beating of the excised heart (or of parts of it) could not be explained in terms of the blood that was no longer streaming in.[28] In reply, Descartes points to the residue of blood in the pores of the heart, which is sufficient for the reactivation of fermentation in the cardiac tissue. And Descartes now turns the tables: how is it conceivable, on the contrary, he asks, that the movement of the separated parts be dependent on a *faculty* of the soul, when the soul is universally considered to be indivisible? Plemp concedes the point in his reply, yet insists that at least the *instruments* of the soul must be present in the excised parts, namely *heat* and *spirit*, which act and animate "in virtue of the soul" (*spiritus scilicet in virtute animae agens*).[29] The Cartesian reply that follows is surely compelling:

> For if these instruments should sometimes suffice on their own to bring about this effect, why not always? And why should we rather imagine that they should work through some power of the soul

> when it is absent, than think that they have no
> need of this power when the soul is present?[30]

Such arguments were all the more convincing since Descartes needed only to remove the plainly unnecessary psychological superstructure from a conceptual framework that was itself already "psycho-materialistic."

Let us now take a look at Descartes's view of the process of the circulation itself. It fits in completely with Cartesian physics, according to which the movement of particles must always occur, in the last analysis, in the form of "vortices."

> I noted above that every place is full of bodies, and that the same portion of matter always takes up the same amount of space. It follows from this that each body can move only in a circle: a body entering a given place expels another, and the expelled body moves on and expels another, and so on, until the body at the end of the sequence enters the place left by the first body at the precise moment when the first body is leaving it (PP II 33, AT VIII-158, CSMK I 237-8 [omitting insertions from the French edition]).

The circular movement is no longer the self-conserving movement of the celestial spheres, but is split up into a series of individual transfers of impulse according to Descartes's third law of nature. The cosmos is filled with fluid vortices of particles with the planets circling about a central star (PP III 30ff., 46ff.); the sun, one of these stars, pushes "the celestial material around it" with force "toward the ecliptic" and carries it along "in a circular movement." (PP III, 84, AT VIII-1 140) It is easy to understand why Descartes unhesitatingly adopted the idea of the circulation of the blood: in his view, it explains the organism by analogy to the automatism of the heavens, the mechanics of the cosmos. The circulation of the blood is no autonomous whole of movement, but a chain of forced individual movements, a "vortex."

> ... the blood that is in the arteries is pushed out of the heart with an effort and by many small jolts to the extremities (TH, AT XI 126).[31]

This pressure or impact continues in the veins as well, supported in addition by the valves and their elasticity.[32] Far from having its own spontaneous tendency or natural movement, the blood, as passive material, is subject to forces of impact according to natural laws.

However, since these forces have a stronger effect in the arteries, what happens there is that particles of blood escape through pores of various shapes in the smallest vessels. In this way, the constant loss of material through the skin ("insensible perspiration") is compensated for (TH, AT XI 126, DCH, AT XI, 245f.). Like the formation of the blood itself, the constitution of the tissues from the blood takes place in a physical manner; the traditional causal principles, on the other hand, are eliminated through the Cartesian identification of soul with consciousness:

> For to suppose in each part of the body faculties that choose, and that attract the particles of food that are appropriate for them, is to invent incomprehensible chimeras, and to attribute to these chimeras more intelligence than our soul itself possesses, since it knows in no way what they would have to know (DCH, AT XI 250-251).

To summarize: we conclude that the universal, vortical motion of particles characteristic of the Cartesian cosmos persists throughout the organism without being subject to any other sort of lawfulness. This motion is merely transformed in the physico-chemical reaction of the central cardiac motor. It now takes the place of the *self-movement* of the body, or of the *soul*, which, in Galenism, already moves the body, so the speak, externally: as a non-corporeal agency. Innate heat is no longer the first instrument of the soul—it replaces the soul; a purely inorganic chain reaction becomes the principle of life.[33]

In all this, the heart itself is only a passive organ carrying out orders; it possesses neither contractility nor sensitivity as presuppositions of a reciprocal relation with the blood. The comparison with clockwork makes it clear that it is not a free rhythm, but a mechanical beat that governs cardiac action. But Descartes also interprets Harvey's doctrine of the circulation in his sense. The circulation no longer appears under the aspect of a totality of movement that represents a cycle of qualitative regeneration and supports, as it were, an autonomous level of the living. Instead, it is decomposed into single particles and instants of movement in order, in this way, to serve as "fuel" for the central motor and at the same time as a "driving belt," which apportions its movement and transmits it to the whole body. Let us now look at this mechanism of transmission.

2. The Movement of the Spirits. At first glance, it seems strange that precisely the tradition-laden and intangible material of the spirits should be a central element in Cartesian physiology. After all, even Harvey, although on principle he is hostile neither to the tradition nor to speculation, considers the spirits dubious and prefers to let them dissolve in the concrete, visible material of the blood. But it is precisely in its speculative character that the stuff of the spirits joins hands with the atomistic basis of modern science. The smallest parts of matter, too, were in the first instance only *entia rationis* with nothing that could be shown to correspond to them in experience. So for Descartes, the invisible, omnipresent spirits, the decisive connecting link in the traditional soul-body hierarchy, presented the appropriate element, in order, once they were reinterpreted as a stream of particles, to act as an *independent agent* moving the bodily machine—and that both in the sense of *steering* and of the *transmission of energy*.[34]

To be sure, Descartes reduces the three familiar kinds of spirits to a single one, the *animal spirits*.[35] According to Galen, these arise from the vital spirits of the arterial blood through a "cooking" in the brain (i.e. in the "rete mirabilis") of the cerebral ventricles; they reach the muscles and sense organs through the nerves and deliver "commands" to move as well as producing sensations.[36] For Descartes, the animal spirits are now only the finest and most rapidly moving particles of blood, which arise from the reaction in the heart and, according to the second law of nature, arrive at the brain along rectilinear paths:

> For according to the laws of mechanics, which are identical with the laws of nature . . . the weakest and least agitated [particles of blood] must be pushed aside by the strongest, which thus arrive at that place on their own (Disc. 5, AT VI 54-5, CSMK I 139).[37]

In their detour from the circulation, they undergo no qualitative transformation, but are filtered through the pores of the pineal gland (epiphysis), a freely moveable cone hanging in the cavity of the brain, which Descartes designates as the seat of the soul.[38] The pineal gland functions as a distributor: as the spirits stream out, it guides them in each case to particular nerves that issue in the inner surface of the ventricles. In this way, the corresponding bodily movement is released—in analogy to the way that the hydraulic automata of the gardens of Versailles are operated (TH, AT XI 130).

The spirits deploy their "energetic" side at the periphery. As the heart is inflated by the blood, so the muscles are inflated by the spirits streaming into them through the nerves and thus contract (TH, AT XI 130ff.). Even at rest, the muscles are filled with spirits; hence, an additional nerve impulse suffices, by means of (hypothetical) nerve valves, to direct the peripheral spirits into one of two antagonistic muscles and thus to bring about its contraction while the other one relaxes (TH, AT XI 132 ff., PA I, 11, AT XI 335-6).

However, with certain arrangements of the muscles, an alternating activation of the valves of the antagonistic muscles comes about even without a central impulse. The oscillating mass of spirits then evokes automatically alternating movements as in breathing or walking (TH, AT XI 139 ff., 196 ff.). A mechanism analogous to the clock pendulum here takes the place of Harvey's vicissitudo in explaining the interrelated sensitivity and reaction of antagonistic muscles.

With this account, we have a brief description of

> ... what structure the nerves and muscles of the human body must have in order to make the animal spirits inside them strong enough to move its limbs—*as when we see severed heads continue to move about and bite the earth although they are no longer alive.* (Disc. 5, AT VI 55, CSMK I 139; author's italics.)

For Harvey as for Aristotle, a severed hand is a hand "in name only". In contrast, the peripheral *spirit*-mechanisms permit Descartes to dissociate the movements of particular parts from the unity of the organism and to see them as automatisms that no longer differ in principle from the movement of separated parts, but are distinguished only by an additional central release-mechanism.[39]

On this basis, Descartes develops a prototype physiology of *reflexes*.[40] According to his model, the nerve openings in the cerebral ventricles are again closed by valves, which open only through pulling on the fibers of the nerve marrow. These fibers are imbedded in the tubes of the nerves and stretch from the valves in the brain to the sense organs or to the muscles in the periphery; in this way, the motor nerves also serve sensory functions.[41] Then when the endings of the marrow fibers in the sense organs are moved through external stimuli

> ... however little that may be, they pull at the same time on the parts of the brain from which they come and at the same time open the entrances of certain pores which are on the inner surface of the brain. In this way the animal spirits ... begin to take their course and go ... to the muscles" (TH, AT XI 141).

Thus a movement against the eyes triggers the muscle of the eyelid to close (PA I, 13, AT XI 338-9), or an injury to the skin causes the withdrawal-reaction of the leg (TH, AT XI 141-2), and this even against our will, that is, in an automatic reflex arc without the cooperation of the soul.

In this way, all unconscious or involuntary movement " . . . depends solely on the arrangement of our limbs and on the route which the spirits produced by the heat of the heart follow naturally in the brain, nerves and muscles" (PA I, 16, AT XI 341-2, CSMK I 335). This "coordination" is of a purely mechanical nature. The body moves like a marionette that activates its own strings; only in voluntary movements does the soul (that is, the mind) enter from without as puppeteer.[42]

These movements occur when the soul steers the spirits in various directions by tilting the pineal gland so that they open the valves in the nerves corresponding to the muscles. Similarly, in the other direction, external stimuli transfer a batch of spirit movements through the nerves to the pineal gland, which evoke sensory perceptions in the soul. To give an account of these interconnections would take us too far from our present theme. However, the mechanisms of the "internal sensations" that have their origin in the body itself are of significance for us here: namely, pain, hunger, thirst, but above all the *emotions*.

Like Harvey, Descartes connects the life of feeling with the circulation. For Harvey, indeed, the blood in its rhythm and movement is the direct *expression* of psychological processes, while in Descartes's view, on the contrary, feelings *are produced in the first place* only through a combination of peripheral and cerebral processes. The "passions of the soul" are only *by-products* of special movements of particles in the blood and nerves.

> It can also be proved that the nature of our mind is such that the mere occurrence of certain motions in the body can stimulate it to have all manner of thoughts which have no likeness to the movement in question. This is especially true of the confused thoughts we call sensations or feelings (PP IV 197, AT VIII-1, 320; CSMK I 284).[43]

Since the *animal spirits* arise originally from the blood or from the reaction of the heart, the state of the circulation also influences the character of their motion. In different amounts and in different degrees of liveliness and turbulence, they stream into the brain and alter, in specific ways, the position of the pineal gland and their diffusion through it. It is in this way that "all the different humors and natural inclinations that are in us . . . are represented in that machine" (TH, AT XI 166).

> One sees in the case of those who have drunk a lot of wine: the vapours of the wine . . . rapidly . . . rise from the heart to the brain, where they turn into spirits which, being stronger and more abundant than those normally present there, are capable of moving the body in many strange ways. Such an inequality of the spirits may also arise from various conditions of the heart, liver, stomach and spleen and all the other organs that help to produce them. In this connection we must first note certain small nerves embedded in the base of the heart, which serve to enlarge and contract the openings to its cavities, thus causing the blood, according to the strength of its expansion, to produce spirits having various different dispositions (PA I 15, AT XI 340, CSMK I 334).

Two influences on the character of the spirits can be recognized here. One rests on the varying composition of the blood that streams into the heart from the other organs. Thus, Descartes gives a new explanation, not only of the classical four temperaments (TH AT XI 166 ff.), but in particular also of the emotion of love and hate. Via the pineal gland, certain thoughts of the soul steer the stream of spirits to the nerves of the stomach, spleen, and gall bladder so that these organs contribute to an altered composition of the blood and finally evoke the corresponding emotions in the soul (PA II 102f., AT XI 403-5, CSMK I 364). The second, even more important effect comes from differences in the opening of the cardiac valves, which is governed by the cardiac (= vagus) nerves. Through the admission of blood, these valves regulate the dynamic of cardiac reaction, evoke many different kinds of variations in the movement of the spirits, and thus participate in the emergence of most emotions, for example, that of joy:

> Thus, if we imagine ourselves enjoying some good, the act of imagination does not itself contain the feeling of joy, but it causes the spirits to travel from the brain *to the muscles in which these nerves are embedded.* This causes the openings of the heart to expand . . . (PP IV 190, AT XI 317, CSMK I 280-281; author's italics.).

And this "expansion of the heart" in turn effects the joyful sensation while fear or sorrow, for example, arise through a constricted heart (PA II 104f., AT XI 405ff., CSMK I 364f.; TH, AT XI 164-5).

The existence of nerves in the heart was already known to Galen and had been generally recognized since the sixteenth century although their function remained a matter of dispute.[44] In Descartes's case, they have, like all nerves, both motor and, as we shall see, sensory functions—the latter, however, only in connection with the emotions. For what follows, it is nevertheless significant that Descartes here describes for the first time *the control of "the muscles around the heart" by the animal spirits of the cardiac nerves* (PP IV, 190, AT VIII-1 316-318, CSMK I 280-281).

Thus, events formerly ascribed to the soul, insofar as they do not include intellectual content, can be entirely reduced to humoral-neuronal processes. This also explains why we experience such occurrences as *related to the body*:

> As for the opinion of those who think that the soul receives its passions *in the heart*, this is not worth serious consideration, since it is based solely on the fact that the passions make us feel some change in the heart.[45] It is easy to see that the only reason why this change is felt *as occurring in the heart* is that there is a small nerve which descends to it from the brain—in the same way pain is felt as though in the foot by means of nerves in the foot . . . (PA I 33, AT XI 353, CSMK I 340-341; author's italics.).

Since the cardiac nerves that call forth a feeling are also of sensitive nature, their activation permits the soul to have the impression of this feeling as in the heart; yet this impression is just as illusory as the localization of pain, which also arises in the brain and is only "projected" into the foot. Descartes does insist that the soul is "united to all the parts of the body conjointly" (PA I, 30, AT XI, 351, CSMK I 339); indeed, it forms a "substantial union" with the whole organism (Reply to Fourth Objections, AT VII, 228; cf. *Sixth Meditation*, AT VII, pp. 56ff.) However, this does not alter the fact that "the soul is of such a nature that it has no relation to extension" (PA I 30, AT XI 351, CSMK 339). Therefore, it cannot be extended in the body. Our spatial sense of our body, though a primary sensation or experience, must be corrected by the principles of Cartesian science.

After all that, what function remains for the emotions as such?

Through its Cartesian reinterpetation, this question actually becomes a double one: about the function of the bodily processes underlying the emotions and about the meaning of their representation in the form of "feelings" in consciousness. As far as the first question is concerned, the emotions represent a kind of expanded reflex mechanism, which, via the circulation and the spirits, adjusts the whole body to a situation that needs to be overcome. Thus, for example, it can

> ... avoid some evil through the use of force, and, in surmounting it or driving it off—to which the passion of *anger* inclines—the spirits have to be more unequally agitated and stronger than usual (TH AT XI, 194).

Thus, external stimuli not only excite direct reflexes, like that of flight, but, via the pineal gland and cardiac nerves, also those surgings up of the blood, which put the body into the state of preparation for danger demanded in a given case. This humoral mechanism then, again via the pineal gland, produces further accompanying reactions of readiness like facial grimaces or cries of pain (TH AT XI 193), or in other cases reactions like laughter or tears, turning pale or blushing. (TH AT XI 194; PA II 114, CSMK I 368) In this way, in the materialized emotions, Descartes is already describing a complete *circulatory system of regulation*, which includes not only simple reflex arcs, but all the bodily functions.

The function of the *mirroring* of these processes in human consciousness—animals have only the corresponding movements of the spirits and of the pineal gland, but no feelings or sensations (PA I 50, AT XI 368-370, CSMK I 348)—consists in the last analysis in their influence on the will:

> ... the principal effect of all the human passions is that they move and dispose the soul to want the things for which they prepare the body. Thus the feeling of fear moves the soul to want to flee, that of courage to want to fight, and similarly with the others (PA I 40, AT XI 359, CSMK I 343).[46]

Man's ability to feel the emotions serves to bring his will into play as an additional adjusting agent. Emotions are signals proceeding from the body which do indeed play an important role in the way we direct our actions, but nevertheless, without their decoding by science, they tell us just as little about the real world as do the impressions of sense.

As we see, therefore, in Cartesian psychosomatics, the feelings do not represent a primarily psychological occurrence expressing itself through the body, but rather a mirroring of bodily regulative processes that is useful for self-maintenance and survival. Their most important center of control is in the reaction of the heart, regulated by the central nervous system, a reaction through which the spirits obtain their special character. The circulatory system transfers these changes to the brain. Descartes's explanations are not uncomplicated; from an injury to the foot to the grimace of pain or rage, the following stations must be passed: sensory nerves, pineal gland, cardiac nerves, circulation of the blood, spirits, pineal gland, motor nerves. Nevertheless, the Cartesian models correspond entirely to modern explanations of psychological reactions in terms of blood pressure, triggering of adrenalin and of neuro-transmission, and so on. In both cases it is a question of reflex mechanisms that respond to stimulation, or of biochemical chains of reaction that evoke the experience of feelings in the brain; and in both cases, the processes of the heart and circulation remain wholly on the side of centrally regulated mechanisms.[47]

This concludes the presentation of the most important elements of Cartesian physiology that are based on the circulation.

d. CONCLUSION

Descartes's theories of the human body, summarized in his *Treatise On Man* as well as in the *Description of the Human Body*, but set out at various lengths in almost all his works, doubtless represent the first projection of a modern, post-Galenic physiology—even though, in altered form, numerous traditional elements re-occur there. On the basis of the circulation of the blood, coupled with the processes of metabolism and assimilation, Descartes forms a new synthesis of the humoral and neuronal systems with the help of which he explains all the essential vital and psychological functions. Despite the speculative character of his theories, which are often developed only as models, they anticipate a surprising number of modern positions. Indeed, precisely as models, Descartes's physiological expositions are still influential today.[48]

It is essential to this approach that the movements and processes in the body should not be subject to a lawfulness specific to the living, but find their explanation in their physico-chemical substructure, which is governed by the same determinants and laws as inorganic nature in general. Just as significant as this subordination of the organism to Cartesian physics is the character of that physics itself. The exclusion of specific tendencies and special directions characteristic of the elements or of effects of impetus peculiar to them; the decomposition of matter and movement into least parts; the replacement of the self-sustaining circular movement of the heavens by the propagation of impact in a cosmic system of vortices—in the period that follows, all this influences physiology as well, in particular research on the circulation, which becomes increasingly physicalistic.

In the other direction, when the organism is integrated into a physico-chemical unified science, the mechanical loses its integrated place in the vital context. Aristotle, Galen, and Harvey also compare particular physiological processes to mechanical devices—but the two are never identified. Now, instead of being an element in a larger process, the mechanical becomes the bearer of the process as a whole. In place of the autonomy of the living being, we have the automatism of its body and thus a radically new perspective on human physiology as well. For now, it is only the phenomena of consciousness that are of non-physical nature (although they, too, are largely physical in their origin). Everything else: unconscious sensitivity or movement, metabolism, growth, and so on, must find their explanation in the material structures involved in each case.

Hence, the basis of physiological processes, innate heat, can neither be (in Galen's sense) the "instrument of the soul," nor (as for Harvey) a vital process specific to the blood, but only an inorganic reaction. At the same time, indeed, the problem of the source of metabolic energy now really becomes a problem. *What moves the organism*—which, now that movement is only *being*-moved, must in the last analysis be pushed from outside? Until now, such a question would have been automatically answered by reference to innate heat. Despite its reliance on nutriment and on air, vital heat was precisely the *autonomous moving principle* of life. Now, with the physical reinterpretation of the innate heat, the blood becomes above all the "fuel" of physiological processes.[49]

Thus, the heart is only the container in whose interior or within whose walls the energy of propulsion is liberated and transferred to the movement of blood and spirits. It is central regulation via the cardiac nerves that determines the beat of this motor—nerves of which, characteristically, Harvey never spoke. Blood and heart now form no balanced and sensitive equilibrium, but are subject to the control of the central nervous system through feeback mechanisms. *With its attachment to the nervous system, the circulation becomes a feedback control system.*

In a body that is soulless as well as lifeless (in which, in other words, the organs are not "as it were themselves living beings"), what becomes decisive is *regulation*, the distribution of the energy released by the reaction of the heart. Hence, the central importance of the nervous system for this kind of physiology. Here, the nerves no longer serve for the perception by the "conductor" of basically autonomous actions by the organs; instead, movements and processes in the body come about only through them. This development becomes evident above all in the concept of the spirits. In Fernel, they were the necessary connecting link between the soul and bodily functions. Harvey identifies them with the blood as the *spontaneously animated* "primary organ" and in the *De Motu Locali* also with muscle in its spontaneous movement; the *spirits* become synonymous with the vitality of the tissues. Descartes, on the other hand, makes them the independent agent of movement, the material substrate in which inorganically produced energy is distributed to all the limbs through the circulatory system and the brain. If Descartes, apparently in the same way as Harvey, calls the blood the 'soul' of otherwise soulless animals—even citing the same biblical passage—this indicates only the reinterpretation of the organism as a feedback system of a purely physical nature:

> . . . I would not wish to say that motion is the soul of animals. . . . I would prefer to say with Holy Scripture . . . that blood is their soul, for blood is a fluid body in very rapid motion, and its more rarefied parts are called spirits. It is these which move the whole mechanism of the body as they flow continuously from the arteries through the brain into the nerves and muscles.[50]

The central distribution of the spirits and the dominant position of brain and nervous system have the consequence *that the traditional distinction between 'natural movements' (dependent on the heart) and 'animal movements' (dependent on the brain) collapses*—a consequence significant for the further development of the theory. That is to say, Descartes also explains intestinal and other unconscious muscular movements on the model of the activity of breathing (TH, AT XI, 140 f.). They all occur in the same way, namely, through the streaming in of the spirits through the nerves, and in the case of voluntary movement, only the triggering by consciousness is added. As we have seen, Harvey, too, in fact stresses the difference between these two movements, but in precisely the opposite direction: namely, insofar as he ascribes them all ultimately to the primary tendency of all muscle tissues to contract.

With Descartes's dismissal of self-motion, however, all movements equally assume the character of reflexes, mechanically pre-arranged programs of movement. What differs now is only their release mechanisms: external and internal stimuli or acts of will.

The only exception left is cardiac movement itself—although it too fails to be really self-moving. Indeed, even here, central regulation is in effect—precisely in Descartes's new idea of a stream of spirit that governs the opening of the valves via the muscle of the heart. This will also play an important role in what follows. With Descartes it means above all that cardiac movements *as causes of feelings* are determined by the central nervous system in such a way that qualitatively different emotions are based on degrees of movement (corresponding to the heating of the blood or the intensity of reaction) that differ only in quantity. At the same time, however, this means that the heart can no longer count, strictly speaking, as the central organ of feeling since its position now represents only the field of projection produced by the nerves for the cerebral "passions of the soul." *The pre-idea of a connection between the heart and the life of the soul*, which was expressed in Harvey's case in the modulation of the cardiac rhythm through "an inner, psychological principle," *is reinterpreted by Descartes as a mechanical connection between heart and brain.*

It is precisely through the double mechanism of the *triggering* of feelings in the body and their *projection back* to the place of triggering that the separation of body and soul first becomes absolute. To carry through this separation, to enable the soul to distance itself from its feelings, and so also to withdraw consciously from its body—that is in the last analysis the goal of ethics as Descartes projects it in the *Passions of the Soul*.[51]

II. THE MOTION OF THE HEART AND BLOOD AFTER DESCARTES

Summary. After consideration of the vital and mechanistic aspects, we now turn to their reception and the history of their influence. Here it becomes evident that the establishment of the circulation as a scientific fact (in the sense of Fleck) cannot be separated from the influence of Descartes. As against vitalistic or magical-hermetical possibilities of interpretation, the mechanical-atomistic reading of the circulation prevailed and opened a horizon of corresponding physiological investigation. It is not only particular elements of Cartesian physiology that recur in most of the sources of the succeeding period, but its fundamental assumptions. Thus, for example, Descartes's abolition of the autonomy of the organs is reflected in the theory of the external regulation of the heart by the central nervous system; and his equation of organic and inorganic processes makes possible the physico-chemical investigation of "vital heat." The machine model of the organism becomes the theoretical foundation for iatrophysical as well as iatrochemical concepts.

In order to demonstrate this development in relation to Harvey's discovery, the essential physiological authors up to 1700 are examined for their points of view and their expositions of the function of the heart and circulation.

(a) In Holland, the Cartesian tradition begins with Henricus Regius and Cornelis van Hoghelande. Regius's theory of cardiac action as driven by the spirits, which was developed from Descartes's theory of the passions, is adopted and further disseminated by the iatrochemist Franciscus Sylvius. The later Cartesians, Theodor Craanen, Cornelis Bontekoe and Stephen Blancaard complete and systematize this tradition in terms of the doctrine of the organism as a hydraulically driven apparatus.

(b) In England, the Cartesian influence leads to diverse conceptions, which do indeed display the effect of Harvey's experimental methodology, but no longer his vitalistic style of thought. In a series of early authors (for example, George Ent, Nathaniel Highmore), despite their admiration of Harvey, the change in their interpretation of the circulation becomes apparent. In Thomas Willis, Richard Lower, and John Mayow, the tradition of Oxford physiology reaches its peak, a tradition that unites the movement of the heart, respiration, and the distribution of the spirits in a comprehensive mechanical-chemical system.

In the third section of this chapter, it will be shown that, despite this dominance of the mechanistic aspect, Harveyan ideas continue to exert an influence in latent form—not only with more vitalistically oriented writers like Thomas Bartholinus, Hermann Conring, or Francis Glisson, but also among avowed mechanists like Jan de Wale, Walter Charleton, Giovanni Borelli, and Johannes Bohn. It is the interpretation of the movement of the heart, in particular, in terms of its exclusively mechanical aspect that is unable to give satisfaction. Thus, we come to "mixed" theories of cardiac action, in which Harvey's stimulus-response model once again finds a place. But in this way, Harvey's idea of an elementary capacity of organic tissue to perceive stimuli is likewise still present—a necessary condition for the rebirth of vitalism in the eighteenth century.

Although the *Treatise on Man* appeared only posthumously in Latin translation in 1662 and in French, together with the *Description of the Human Body*, in 1664, Descartes's physiological ideas had been generally known since the late thirties of the century through his other works and through the versions disseminated by his disciple Henricus Regius (1598-1679), and their influence in scientific circles can hardly be overestimated. Apart from the Dutch Cartesian school, Descartes's teachings also found especially fruitful soil in England and

Scandinavia—in other words, in the countries that were at that time scientifically most advanced, also the first in which Harvey's discovery was able to establish itself. Finally, in the second half of the seventeenth century, Cartesianism, partly in alliance with a renewed atomism of a Gassendist cast, rose to become the dominant direction of European science, and it remained definitive for the image of nature and even more of humanity itself, even when mathematico-physical unified science in the form envisaged by Descartes proved impracticable.[52]

The Cartesian influence also extended to physiological views about the movement of the heart and blood. The comparatively rapid acceptance of the doctrine of the circulation was inconceivable without the authority of Descartes and his philosophy. Only that philosophy promised to assimilate Harvey's revolutionary discovery into a new physiological system, and at the same time, offered a defense against contemporary forms of skepticism especially threatening to a science in upheaval.[53] It was the *Discourse on Method* (1637), one of the most widely read works of the time, that first redirected attention to Harvey's treatise of 1628, which had been widely ignored or rejected. Consequently, the Cartesian point of view assisted in the establishment of Harvey's discovery, but at the price of influencing the way it was conceived. Harvey's vitalistic tendencies, suggested in the *De Motu Cordis* and constitutive for his later writings, remained largely unnoticed and, with a few exceptions, were not pursued further. Instead, Harvey appeared as the protagonist of an "experimental science" consciously renouncing "occult" or soul-like agencies, for which only the corpuscular philosophy provided the real foundation.[54]

Descartes himself, though a medical amateur, counted as an authority in the physiological no less than in the philosophical sphere, one whose opinion even eminent physicians accepted.[55] His immediate influence can be measured above all by the fact that his "explosion theory" of cardiac action found numerous adherents—by no means only among the Cartesians: e.g. in Henricus Regius, Cornelis van Hoghelande, Franciscus Sylvius, Theodor Craanen or George Ent and Thomas Willis in England (see below). As we have seen, Descartes's theory reversed systole and diastole, though agreeing with the Galenic tradition in making diastole the active part of the cycle, and in addition, he had some difficulty in explaining the continued beating of the excised heart, postulating fermenting remains of blood in the pores of the inner wall of the heart. Although on closer scrutiny this theory proved untenable, it nevertheless satisfied the demand Descartes had made, in opposition to Harvey, for the elimination of vital or psychological elements from cardiac action. It was precisely this that made Descartes's explanation seem so attractive; and even though its untenability gradually became evident, Harvey would in no way have been able to rejoice at his own victory. Instead, when the Cartesian theory failed, people just looked for other mechanistic solutions. In any case, discussion of the Cartesian theory of the heart counted among the obligatory components of physiological writings on the circulation up to the end of the century.[56]

For the most part, we have to look for Descartes's influence on his successors in their physiological *thought-style* rather than in individual explanations of particular points. The important Danish anatomist Nicolaus Steno (1638-1686), himself a thoroughly critical admirer of Descartes, saw this very accurately as early as 1666:

> No one but he has explained mechanically all the actions of man, and especially those of the brain; others describe for us man himself; M. Des Cartes speaks to us only of a machine, which, however, makes us see the inadequacy of the others' doctrines, and teaches us a method of investigating the functions of the other parts of the human body, with the same evidence with which he demonstrates to us the parts of his man-machine, something no one has done before him. Thus we should not condemn M. Des Cartes if his system . . . does not prove to be entirely in agreement with experience; the excellence of his mind, which appears chiefly in his *Treatise on Man*, excuses the errors of his hypotheses.[57]

Before we proceed to the study of particular writers, let us first consider in general some tendencies in physiological thought after Descartes. In physiology, the withdrawal of the soul from the body is gradually completed. The soul's place is taken, not by a vitality specific to individual organs such as Harvey envisaged, but by centrally driven and controlled functions that are describable in physico-chemical terms. Not that the soul was wholly banned from physiological theories—but it now appears, reduced to the intellectual soul in the brain, only as the final link in the chain of physical processes in the body.[58]

However, these processes now need their own driving mechanism, a *motor* in the physico-chemical sense. In this context, Descartes's identification of the energy-giving and blood-moving process does indeed prove untenable. The conception of the heart as a pump prevails over the notion of an "explosion motor." Especially after the exact measurements of temperature by Lower (1669) and Borelli (1689), the heart must now withdraw definitively as the seat of central heat. In the last analysis, it is the circulation, unifying all the phenomena, that ousts it from this position.

But then where does the energy for bodily processes come from? Thought about this moves from the heart to the blood—apparently as with Harvey, but not at all in his sense. The blood is conceived as chemically analysable material composed of inorganic particles, and thus as the carrier of an energy-producing reaction, conceived chiefly in terms of fermentation. *Breathing* is now also assimilated to the inorganically reinterpreted problem of vital heat: without the fire in the heart, it can no longer serve as cooling agent. It turns out that the change in color of the blood takes place, not at all in the heart, but in the lungs (Lower, 1669). Attached to this discovery are the first theories of a union with the blood of the inspired air, understood as corpuscular, and of the necessity of that union for the chemical reaction of combustion and for the origin of the animal spirits in the nervous system (Mayow, 1674). Thus, the movements of the body become wholly the product of the external, universal movement of particles, which propagates itself via the air in blood, nerves, and muscles.

Granted, Descartes's oversimplified theory of the mechanical inflation of the muscles, produced by the spirits, is subsequently replaced by different conceptions—fermentative reactions in which the blood participates, triggered by the streaming in of the spirits. Nevertheless, the *principle* of Cartesian physiology prevails extensively: the central causation of all movements, that is, the abandonment of "natural motion" and of autonomous muscular movement in favor of "animal motions" governed by the spirits. Here, then, we also find the solution of the problem of how cardiac action could still be explained in a mechanistic fashion after the collapse of the "explosion theory"—a solution that is already foreshadowed in Descartes in the connection between the cardiac nerves and the emotions. Cardiac action is not understood as perception and response, but *is also tied to the central distribution of the spirits, which set the beat for its motion.* Like all muscles and organs, the heart is now moved *from outside.*

It is not only the exclusion of autonomous muscular movement that makes a basically autonomous reciprocity of heart and blood impossible. It is also precisely the mechanistic turn in embryology that prevents a continuation of Harvey's epigenetic idea of an elementary movement of the blood as the presupposition of cardiac action. In preformation, the flow of blood necessarily becomes a secondary phenomenon, resulting from the preexistent organic structures.⁵⁹ The mechanical pumping action of the heart then also dominates the physiology of the developed circulation, which is viewed as a merely passive movement, reacting to pressure. Only much later does the question arise, whether the force of the heart is really adequate to drive the total mass of blood through the ramifications of the stream bed.

In these circumstances it seems more interesting to investigate in the various subsequent writers the action of the heart itself, and thus the critical point in expositions of the circulation, which otherwise appear relatively free of problems. It will turn out that here, as against the dominant mechanistic tendencies especially in Holland and England, "vitalistic" views could yet be maintained, views that could later be taken up again. Admittedly, our investigation must be selective. I have therefore tried to take account of the more significant writers up to 1700, insofar as they expressed themselves on this subject.⁶⁰

a. HOLLAND

1. Henricus Regius (1598-1697). Starting in 1638, Regius taught in Utrecht as the first Cartesian professor of medicine and from then on was also personally acquainted with Descartes and friendly with him.⁶¹ He sought his own synthesis of Cartesian philosophy with physiology and medicine. The first result was the *Physiologia sive Cognitio Sanitatis* of 1640, followed by the *Fundamenta Physices* in 1646. Both were written without knowledge of Descartes's *Treatise on Man*—Regius read it only while the *Fundamenta* was in press—but with cognizance of the most important elements of Cartesian physiology.⁶² Finally, the *Fundamenta Medica* (1647) represents a first application of these principles to clinical medicine.

Regius has special significance for the dissemination both of Cartesianism and of Harvey's doctrine of the circulation—not least in England. His writings were much better known than Descartes's views on physiology, which were published only in scattered fashion. In his publications from 1640 on, Regius repeatedly took Harvey's side, especially in the controversy with James Primrose (1598-1658), who was at that time the English physician's most outspoken opponent. Regius's—extensively Cartesian—theory of cardiac activity even found its way into the first English language edition of the *De Motu Cordis* in 1653.⁶³ His influence as one of the leading Cartesians in Holland persisted even after Descartes had broken with him in 1645 because of his materialistic tendencies and had publicly distanced himself from his former friend in 1647.[64]

In the *Fundamenta Physices,* Regius develops the Cartesian doctrine of matter and motion in its essential points (chapters 1-8); he reproduces the first and second laws of nature and demonstrates the necessity of the circular propagation of the movement of particles, in other words, the theory of vortices.[65] According to Regius's modern sounding definition, in this physical world, living bodies are those whose parts are so arranged that they maintain themselves in their substance through a constant supply of freshly assimilated particles. At the same time, their "vegetative soul" consists in this arrangement (*Corpora viva sunt, quorum partes ita sunt temperatae et conformatae, ut corporea eorum substantia, quae perpetuo dissipatur, per succum praeparatum et in interiora impulsum . . . conservetur . . . Haec partium dispositio est istorum corporum anima vegetativa;* p. 143). Perception and movement come about "as in automata" (*eodem modo ut automata agitantur et moventur,* p. 153). This is put even more clearly in the *Fundamenta Medica:*

> Those animal movements are automatic which, without the participation of the soul or mind, solely through the movement of the corresponding organs (that is, of the spirits, the nerves, the brain and the muscles, stimulated by an internal or an external object), . . . are executed by human beings as by an automaton; but they can also be performed by the mind, if it turns its attention to them.[66]

The animal spirits are separated from the blood in the brain (*in ventriculos cerebri a corde propulsus,* p. 159); the respired air, which mixes with the blood in the lungs, is also again involved here (p. 224). The account of the expansion of the muscles, of the peripheral valves in the nerves, and of the alternating involuntary motions closely follows the Cartesian pattern (p. 231, 235ff.). But Regius also speaks of a "circulation" of the spirits, which return via the veins to the heart (p. 225rf.), a thought that does not yet occur in Descartes.

Thus, the point of origin of the spirits is ultimately the heart, where the "fire without light," the process of fermentation, physically interpreted, serves as the triggering mechanism of the body. Regius subscribes to Descartes's theory of the heart, although with an important addition: diastole (and thus the streaming out of the blood) is effected, not only through the expansion of the blood, but also through the streaming of the spirits into the fibers of the heart (p. 179). This comes about, Regius declares, from the influence of the passions on cardiac activity, and this too, in the same way, must be attributed to the spirits (*spirituum in cordis fibras influxus patet ex cordis, a spiritibus, in animi pathematis, perturbatio*). Regius goes even further and points to an alternating mechanism of the spirits to account for the systole and diastole of the heart, analogously to the movement of breathing or of antagonistic muscles; in this way, the beating of the isolated heart would also be explicable:

> But the spirits flow alternately into the (different) fibers of the intact or also of the excised and dissected heart, in the same way in which we see the spirits flow now into these, now into those fibers and muscles in the amputated tail of living lizards and in animals breathing in their sleep.[67]

Descartes's *Principles* had been published two years earlier in 1644, and Regius surely knew the views expressed there about the influence of the spirits on the heart. From this, however, he inferred a continuous effect and thus was the first to associate the cause of cardiac activity itself with the mechanically conceived animal spirits—as we shall see, a development of Cartesian ideas that would be rich in consequences.[68] An important reason why Regius found himself forced to extend Descartes's doctrine in this way was the increasing criticism of the Cartesian theory of the heart.[69] The activity of the heart was now doubly protected, as it were, through the fermentation of the blood and through the streaming in of the spirits; and "there is no reason to attribute to the heart any force, be it one of attraction, or magnetism, or a contribution to nutrition, or anything else unintelligible and superfluous."[70]

Regius's account of the circulation of the blood also corresponds to the mechanical theory of the heart (*sanguinis circulatio est motus, quo sanguis a corde per universi corporis arterias propellitur in venas . . . et e venis porro repellitur in cor;* p. 185); it adds nothing to Descartes's view; thus we may turn now to the next author in the Cartesian tradition.

2. Cornelis van Hoghelande (1590-1651). The *Oeconomia animalis* of the Leyden physician Cornelis Hoghelande, who was also a friend of Descartes's, belongs, along with the work of Regius, to the earliest Cartesian physiologies. It appears as a long dissertation in his Cogitationes, quibus dei existentia; item animae spiritalitas, et possibilis cum corpore unio, demonstratur (1646); thus, it is an essential part of a philosophical treatise in which Cartesian mind-body dualism is explained.

Hoghelande summarizes the principles of his physiology toward the end of the essay. He turns repeatedly against pseudo-explanations through "forms, faculties, antipathies or sympathies, the foresight of nature, archai, attraction, occult qualities, or other dark principles of similar value."

> ... we are of the opinion, that all bodies, however they act, are to be viewed as machines, and their actions and effects ... are to be explained only in accordance with mechanical principles.[71]

This holds also for the effects of bodies on one another as well as for the "natural" movements proceeding from living bodies themselves. For in the last analysis, "no body moves on its own, but is always moved by another" (*nullum corpus a se ipso, sed ab alio moveri*, p. 30). The apparently autonomous movements of animals are in fact necessitated by the external movement of particles ("stimuli" on the sense organs), which are transmitted through the animal spirits.[72] Only in man can a mental influence precede this; otherwise, man too is subject to the automatism of movement in its pure form.[73] Thus, life can be defined as the "continual distribution of the spirits from the brain through the nerves, which affords the body sensation and movement" (*unde vita animalis ... definietur: continua spirituum, sensum motumque corpori exhibentium, ex cerebro per nervos distributio;* p. 126).

All this resembles the contemporary physiology of Regius as does the reinterpretation of the vital heat of heart and blood. This consists in a chemical reaction analogous to that of spirits of niter or oil of antimony; that is, in contrarieties of particulate movement, which may be called "antipathies" only metaphorically (*internorum motuum diversitate ... aliis forte antipathia dicta,* p. 30). Here, Hoghelande is making an effort to free chemistry of "incomprehensible and occult qualities" (loc. cit.). Finally, he attributes these exothermic reactions to the influence of the particles of Cartesian ether, that is, of the universal "subtle matter" that God set in motion at the beginning (loc. cit.). This subtle matter, which flows through all bodies and moves them, is really what others call the "world soul" (*materiam subtilem ... esse instar Animae Mundi,* p. 97). Through the inspired air, the spirits too share in this material. We will meet this notion again in Theodor Craanen and Friedrich Hoffmann.

Most of Hogheland's treatise is devoted to the physiology of the heart and circulation. The movement of the blood is explained purely mechanically, and Harvey is praised for his exposition "*secundum manifestas mechanicae leges*", according to the manifest laws of mechanics (pp. 82ff.). In his defense of the Cartesian theory of the heart, Hoghelande too attempts to undertake an adaptation. But unlike Regius, he offers an additional hypothesis for the contraction of the heart: "if there really is such a thing" (*siquidem vera fiat constrictio,* p. 61). Instead of seeking his refuge in pores inside the heart as Descartes had done, Hoghelande simply lets fermentation take effect for a second time from without—namely, in the blood of the coronary artery:

> But if any one asks, further, through what means that convulsive movement is maintained so long in excised parts of the heart, ... then, in the light of what has previously been demonstrated, namely that the parenchyma of the heart has in a high degree a fermentative action, he will take note of the entry of blood into this parenchyma through the coronary artery.[74]

The heat of fermentation, that is, the inner movement of particles in the blood, makes the cardiac fibers contract (*quo moto ... [particulae sanguinis] in fibras ... cordis impingentes, motum quendam convulsivum contractione fibrarum excitant;* p. 60f.). To demonstrate this, Hoghelande refers to an experiment with the heart of an eel, which he had pushed under water when empty of blood. Only after the infusion of blood did it once more begin to move.[75] In this way, therefore, the convulsive movement of the isolated heart would be explained. The heart's action *in situ*, on the other hand, has two different causes: systolic (external) and diastolic (internal) fermentation—whereby Hoghelande continues to consider diastole, in which the blood is supposed to leave the heart, as primary. Here we have, then, already two essential variants for subsequent theories of the heart. Now, even among declared Cartesians, the conception of the pump gradually prevails over the original Cartesian view; above all, it demands a mechanical explanation of systole, that is, of the contraction of the heart.

3. Franciscus Sylvius (François de la Boe, 1614-1672). Although his own position was only neighboring to that of the Cartesians, Franciscus Sylvius represents in a way the connecting link between the earlier and the later Cartesians. From 1638 on, he taught in Leyden (from 1658-1672 as professor), collected around himself many pupils, some of whom became important, and belonged among the most influential Dutch physicians of the seventeenth century.

Sylvius was quick to take a position in favor of Harvey's doctrine of the circulation, though at first only in connection with the Cartesian theory of the heart.[76] From 1640 on, he was also personally acquainted with Descartes. Descartes's influence is evident in many passages in Sylvius's writings (for instance, in the adoption of the theory of the pineal gland, but also in the basic tendency of his doctrine). Along with Thomas Willis in England, Sylvius is the chief exponent of a "mechanized" iatrochemistry, which sought to explain vital processes on the basis of the inorganic and, in a sense, to "distill out" the material content from the ideas of Paracelsus and van Helmont.[77]

For Sylvius, chemical reaction in the forms of (slower) fermentation and (more violent) "effervescence" becomes the motor of all living processes; the basic model for this is provided by the reactions of acid and alkaline agencies. Sylvius's chief interests were in the processes of digestion and in the brain. Here, however, we shall present only his views of the heart and circulation, as they are to be gleaned from his *Disputationes Medicae* (1659 ff.).[78] According to this text, the blood represents a heterogeneous mixture, which undergoes an "effervescence" and dilation in the right ventricle much as it does in the Cartesian doctrine, and is recondensed in the lung through mixture with particles of air in order to react a second time in the left ventricle and reach its "perfection" (III, 6; VII, 74-78). Nevertheless, the Cartesian theory of the heart no longer seems adequate to Sylvius, on the one hand, in view of the beating of the isolated heart, and on the other, because of the fact that only the surplus portion of the dilated blood would leave the cardiac chambers—a thesis that observation contradicts (IX, 43 f.). According to Sylvius's compromise solution, the heart manages to anticipate the exodus of the swelling blood through its contraction as mediated by the spirits:

> ... (concerning the cause of the circulation) we are of the opinion that the blood is kindled by the inner fire in the chambers of the heart and made thinner, and thus reaches its perfection, so that it takes up more and more space and consequently extends the walls of the heart further and further, until finally, since the chambers can no longer expand any further, (the blood) would break out into the arteries by itself ... —were it not that the parenchyma of the heart, burdened by this dilation, calls to its aid the spirits, which stream in in sufficient quantity to make the muscles of the parenchyma contract and thus to drive the blood . . . into the arteries.[79]

Thus, the contraction of the heart is evoked by the spirits, which Sylvius, following Descartes, reduces to one kind, the animal spirits. They are produced from the blood in the brain through a process of filtration and chemical reaction, distributed through the nerves, and return back to the blood in a circular flow, namely as lymph—it is in this way that Sylvius interprets the new vessels described by Aselli in 1627 (IV, 29, 31). Along with their various psychological functions, these new vessels produce "animalistic," that is, muscular, movements as well as the movements of peristalsis in the hollow organs (IV, 39).

It is instructive for the development of the original Galenic conception, to see what now counts for Sylvius as "natural motion", namely, the dilation of the heart through the reaction in its interior, as the only movement still not evoked by the spirits:

> Natural, in our view, should be called the dilation of the ventricle by the thinning blood, animal, on the other hand, its contraction, which is executed by the muscles and is in a way subject to the will (!).[80]

It is evident that two different conceptualizations are being imposed on one another here: the Galenic, according to which it is the muscular movements mediated by the animal spirits that are the "voluntary" ones, and the Cartesian, according to which the spirits are able to regulate in effect automatically or involuntarily all the movements of the body—and now also those of the heart. Sylvius has clearly not completed this reinterpretation, and allows himself to be led by linguistic usage to the remarkable conclusion of bringing the contraction of the heart into connection with the will.

Thus for Sylvius, too, the central distribution of the spirits is the genuine cause of the circulation of the blood. For "the blood is indeed kindled by the inner fire in the heart, thinned out by it, and thus made more mobile; still, it is moved essentially and driven forward by the parts that contain it."[81] Among these parts are the heart, but also the vessels and especially the vena cava with its peristaltic—that is, spirit-mediated—movement. There is no autonomous movement of the blood except for its chemically induced expansion.

Finally, the regulation of the circulation as well is accomplished essentially through the control of the spirits. The various qualities of the pulse are chiefly to be derived "from the greater or lesser quantity of the animal spirits, which move to the cardiac muscles and contract its parenchyma more violently or more sluggishly"; here too, we find again the original Cartesian thought of a regulative function of the cardiac nerves.[82]

4. Theodor Craanen (1620-90). Theodor Craanen, the successor of Sylvius as professor of medicine in Leyden (1673-89), succeeded in establishing Cartesian physiology firmly there and making it the dominant movement in Dutch medicine until far into the eighteenth century. Both his *Tractatus physico-medicus de Homine* (1689) and his *Oeconomia Animalis* (1703), which we shall consider here, are based entirely on the foundation of Cartesian philosophy. He writes:

> First we shall treat man the automaton, that is, the animal body, and see what functions result from its most ingenious structure, without the invocation of our mind. Second we will inquire into the nature of our mind . . . Third we shall unite this mind with the body.[83]

The location of this union is the pineal gland (pp. 153ff.); it determines the course of the animal spirits, on which all the movements of the body depend (*a solis spiritibus animalibus a cerebro per nervos in istas partes delatis dependisse omnem motum;* p. 149). Muscular contraction corresponds to a dilation by the spirits, regulated by the valves of the nerves (136ff.).

Like Sylvius, Craanen distinguishes between fermentation and effervescence, but traces both back to Descartes's "matter of the first element", that is, to the ether, which penetrates all the pores and represents the real source of movement in the universe (p. 15f., 39). The function of those thermal reactions consists in the formation of the blood, in the refinement of its particles to spirits, and finally, in the motion of the heart.

Here, Craanen shows caution; he does indeed defend Descartes, but he takes note of the controversy that has been taking place about this problematic (*circa hanc rem est disputatio,* p. 46) and he finally decides for a combination of all the defensible mechanistic explanations. The cardiac nerves appear to him too fine to introduce spirits in sufficient quantity so that they alone could achieve a muscular contraction against the swelling blood in the interior of the heart (p. 48). In addition, Craanen therefore introduces a secretion of the coronary arteries, which is supposed to fill the "tubules" of the cardiac parenchyma and, together with the animal spirits, release an effervescence (p. 47).

But, above all, the very swelling of the blood in the ventricles also contributes to the contraction, insofar, namely, as its finer parts penetrate from within into the wall of the heart and dilate it. With this, Descartes proves after all to be right in a fashion; through its dilation in the parenchyma, the blood pushes itself on its own, as it were, out of the heart:

> Thus the blood produces contrary results, it dilates and contracts; it dilates, namely, insofar as it is within the ventricle; and it contracts and narrows the ventricle, but through the part of the spirits that enter the parenchyma of the heart.[84]

As is evident, Craanen has simply synthesized the decisive elements of the theories of Regius, Hoghelande, and Descartes respectively to produce a new mechanistic explanation.

5. Cornelis Bontekoe (1647-1685). Not without consequences, Cornelis Bontekoe, the student of Sylvius and Craanen, carries the tradition we have been examining to an extreme that in many respects seems almost unique. In his posthumously published *Oeconomia animalis* (1688), he develops his physiology on the foundation of a Cartesian physics of particles elaborated with the help of chemical innovations.[85]

Presumably under the influence of Lower and Borelli, Bontekoe, for the first time in the Cartesian tradition, abandons entirely a physico-chemical reaction or specific heat in the heart and concentrates wholly on hydraulic explanations:

> Systole is the primary movement of the heart and depends in no way on the blood, but entirely on the cardiac muscle and the sap of the nerves.[86]

Since Malpighi's *De cerebri cortice dissertatio* (1665) and Borelli's *De motu animalium* (1680-81), the doctrine of the *succus nerveus* as a lymphlike secretion of the blood had gradually replaced the theory of the spirits, without changing anything essential in the principle of the theory.[87] How does this sap effect muscular contraction —including that of the heart? Bontekoe assumes that the tissues of the body are composed essentially of tubular structures.[88] Thus even

> . . . the interior of the muscles consists, not of fibers, but, like everything else, of tubules, nervous tubules coming from the nerves, and arterial tubules from the blood-vessels; if now more nervous sap and blood is driven into these tubules from the heart and brain, what else can happen but the filling, swelling and expansion of the tubules?[89]

Thus, all muscular movements, including that of the heart, depend on nervous sap and the streaming in of blood. What is characteristic of Bontekoe's physiology, however, is the narrow interconnection, indeed, the complete dovetailing of the systems of blood, nerves, and muscles. In order to make possible the multitude of muscular actions, the sap of the nerves must constantly be transferred in the body; it returns through the lymph to the blood, which functions as its "vehicle" (p. 23). The circulation of the blood now serves almost entirely for the circulation of the nerve-sap, which is now coupled with it, and which in its turn drives the circulation (*[sanguis] circuitu debuit circumferri, ut ille succus nunc his nunc illis membris et suppeditari possit, et ab iis ad fontem dimanare;* p. 23). Every movement of the body is equivalent to a displacement of fluid and must be compensated for by the heart:

> Therefore when the muscles are moved, the heart must move in a determinate fashion, in order to send more blood to the muscles and to cause the brain to to move as well and to incite it at the same moment to provide some of the nerve-sap to the muscles.[90]

Even functionally, the physiological compartments cannot be separated from one another. This principle of the simultaneous movement of fluids also determines the circulation of the blood itself, so that the heart "expels the blood from itself and by the same labor pushes new blood in" (*cor et expellat ex sese sanguinem, et eadem opera novum in se impellat;* p. 93). For Bontekoe, the bodily machine becomes a unified system of communicating pipes, a hydraulic Perpetuum Mobile:

> ... thus in the end the heart moves everything, but the heart itself is moved by the nerve-sap ..., so that a wonderful circulation occurs here ... (The heart) pushes the blood with great force to the brain, so that ... the nerve-sap is secreted, and . . . driven into the nerves, among others also into the cardiac nerves ..., so that together with the dilation of the contracted cardiac chambers ... that nerve-sap streams into the flesh of the heart, to (produce) a new contraction. And thus it is that the heart moves everything, and also itself ...[91]

As we can see, Bontekoe has carried out the cancellation of the autonomy of the organs to the farthest consequence; like the gears in a clock, the individual parts and hydraulic circulatory systems fit firmly into one another. Bontekoe himself once makes use of this image to explain the divergent frequencies of the joint activities of the heart and of breathing: namely, as different "translations" like those in the works of a clock.[92]

Corresponding to this image also is the preformationist embryology of the *Oeconomia animalis:* Harvey, Bontekoe tells us, mistakenly assumes a primary movement of the punctum saliens, although it never "moves without the other parts; for they have all originated at the same moment . . . and besides they owe all their motion to the fluids which come to them from the mother."[93] Thus, the first impulse to move comes from outside; but then the body becomes a closed system:

> Human life consists in the so-called union of mind and body; but the mind lives, and the body too lives its own life. The latter does not depend on the mind, and it is nothing but the revolution and the movement of the blood and the fluids.[94]

6. Stephen Blancaard (1658-1702). The physiology of Blancaard, a Cartesian and a student of Craanen at the beginning of the eighteenth century, already shows clearly the influence of more recent authors (Willis, Mayow, Bohn) of whom we will hear more shortly. Granted, for Blancaard, too, the body is primarily a machine, which "moves only like a hydraulic or water clock."[95] However, he takes over from the English physiologists the production of heat, now placed in the lungs, through the "effervescence" of particles of blood and air (p. 193f.). According to Blancaard, the real meaning of this reaction consists in the production of the spirits, which are then separated from the brain as nervous fluid (he uses the two terms as synonymous). Wholly in the Cartesian manner, Blancaard attaches to the spirits a detailed account of mental functions, passions, and the cardiac activity corresponding to them, the mechanism of imitation, and so on (pp. 249ff.).

What is chiefly of interest here is Blancaard's version of muscular and thus also cardiac movement. The structure of the musculature is now entirely determined by the vessels: "the vessels become narrower and narrower at their extremities and finally change into muscles, glands, etc." (*arteriae circa extremitates sensim magis magisque angustiores fiunt, ut tandem in musculos, glandulas, etc. mutantur;* p. 196). Thus, the muscles are composed for the most part of *"vasa minima,"* which are closely connected with the nerve fibers. The streaming in of spirits shortens these fibers, and this leads to the widening of the vessels. The real swelling of the muscles now comes about through the blood—a conception oriented to Borelli (pp. 242ff.).

In contrast to the voluntary muscles, the organic movement of peristalsis and the systole and diastole of the heart depend "only on the mechanical structure of the parts, which are moved in hydraulic fashion by the fluids streaming in."[96] All formerly "natural" movements are thus centrally regulated. Blancaard now solves the problem of cardiac movement by turning diastole as well as systole into active movements and thus explaining both (as Regius had already suggested) on the model of alternating muscular activities:

> ... just as the other limbs have their antagonists, so the heart, intestines and so on have two different fibers inducing movement [what is being referred to here are the *vasa minima;* author]; for when the one set contract, the others open, that is, as soon as the former fill with blood, the others are emptied; this is an alternating movement and is called either peristalsis or systole and diastole.[97]

Basically, what we meet with here, on a mechanistic foundation, is once more the Galenic conception. Galen had assumed dilating longitudinal and contracting cross fibers to account for the active diastole and systole of the heart.[99]

Oddly enough, Blancaard adds to this mechanism of cardiac action an element that is in fact quite alien to it: in addition, the heart is stimulated by the blood; its fibers perceive the stimulus of the blood that is newly streaming in and is thus "forced into more violent swellings and contractions" (*fibrae eius ... stimulum ... novum quoties ab irruente ... sanguine persentiscunt toties in violentiores turgescentias ac contracturas abripiuntur;* p. 196).

Blancaard took this idea, which he does not elaborate further, from Borelli and Bohn. Later, we shall look more closely at this peculiar enlivening of a basically mechanistic view with "vitalistic" elements. But first, we must look for Cartesian influences in the English school of physiology in the second half of the seventeenth century.

b. ENGLAND

In connection with Harvey's discovery, or in other words, after the establishment of the theory of the circulation about 1640-50, there developed in England a tradition of intense physiological research, which crystallized above all in Oxford as its center and which lasted until about 1680. Its representatives considered Harvey the pathbreaking representative of a new medical science and appealed to his experimental methodology. They regarded it as their chief task to place his discovery in the context of a basically revolutionary physiology. In this task, however, they made widespread use of the new continental corpuscular philosophy, both in Cartesian and in Gassendist form—the two tendencies were not always clearly distinguished from one another. In close collaboration with primarily physically, chemically, or mathematically oriented scientists (Robert Boyle, Robert Hooke, Christopher Wren among others), the English physiologists developed new theories, especially about the heat and respiration of the body. Nor did views of the functions of the heart and circulation fail to be influenced by French and Dutch mechanistic thinking. If, therefore, Harvey's discovery represented the starting point of new physiological researches, this by no means meant that these were carried on in his sense.[98]

We shall point this out first in some early writers, and then in the most significant English physiologists of the second half of the century, namely, Thomas Willis, Richard Lower, and John Mayow.

1. Early Writers (ca. 1640-50). It was not until ten years after the publication of the *De Motu Cordis* that a group of chiefly younger disciples of Harvey gradually assembled, seeking to propagate his theory and strengthen its support with their own experiments. Among them was George Ent (1604-1689), who had been friendly with Harvey since 1636 and soon appeared as one of the chief exponents of Harvey's theory of the circulation and of his methodology at the London College of Physicians. His *Apologia pro circulatione sanguinis* appeared in 1641 as the first work on the circulation in England after Harvey's *De Motu Cordis*. Although it was dedicated to Harvey and exactly copied his theory along with its physiological implications, nevertheless the *Apologia* already displays the tendency that was to dominate English physiology. That is: on the question of cardiac motion, Ent decisively sets himself in opposition to his master. "In what ... follows, I shall offer no support to Harvey, since I do not agree with him in this matter."[100] Harvey, he holds, falsely ascribes the ejection of the blood to systole, "since he holds it to be necessary that that movement should be excited by a faculty of the heart" (*cum pro confesso haberet, motum illum a cordis facultate excitari;* ibid.) But it is not permissible to ascribe a "faculty" or a power of contraction to the cardiac parenchyma or to its fibers (*non autem a facultate aliqua fit contractio,* p. 136). This corresponds to the objection of the Cartesians against Harvey, and Ent's own explanation is also precisely that of

Descartes: the blood, expanding through being heated, makes the heart dilate and itself moves out into the aorta:

> ... one may not to ascribe this movement to the cardiac parenchyma, but rather to the blood that is heated in the left ventricle, through whose swelling the ventricle gives way and diastole results.[101]

Diastole is the decisive movement, systole only the secondary relaxation (p. 136); Harvey had confused systole and diastole (*systolen pro diastolen acceperit;* p. 135). The heart does not drive out the blood; rather, the expansion of the blood moves the heart (*sanguinem non expelli a motu cordis, sed hoc moveri ab illius intumescentia;* p. 146). Descartes's theory had been known for four years from his *Discourse on Method;* if Ent does not refer to it, it is nevertheless entirely implausible that he would have invented the same explanation himself—his argument corresponds too exactly to the Cartesian.

Further development in this area now proceeded much as it was doing in the Netherlands. Descartes's theory of cardiac movement proves to be increasingly untenable and attempts are made to replace it by other mechanistic explanations. Thus, for example, in 1644, Kenelm Digby (1603-1665) rejects the Cartesian version in view of the heart's continuing to beat when empty of blood, but he also rejects the traditional vitalistic interpretation (" ... we must inquire after another cause of this primary motion").[102] In his view, the contraction occurs through a shortening of the fibers owing to the blood's moist condition. How far the Cartesian point of view penetrated circulatory physiology in this period is illustrated, for example, by the 1650 lectures of William Petty (1623-87) in Oxford. Petty, who had studied with Hoghelande and had assimilated Descartes's doctrine of nature at least in principle, taught Harvey "à la Cartésienne" with special emphasis on the mechanical nature of cardiac activity and on the corpuscular constitution of the body fluids.[103]

Such tendencies were also evident in Nathaniel Highmore (1613-85), who in 1651 published the first British textbook of anatomy (*Corporis humani disquisitio anatomica*), which was followed by his atomistic embryology.[104] Highmore views the circulation of the blood as a movement that is compelled in a purely mechanical fashion (*sanguinis circulatio est motus in sanguine violentus, a cordis contractione, impetuose illum in arterias impellens ... factus;* p. 146). He interprets its functions (maintenance of the blood, distribution of heat and of material) on a corpuscular basis.[105] The same holds for the spirits, which he, like Descartes, reduces chiefly to the animal spirits and interprets as a stream of particles (*unicam ergo spirituum causam materialem assignare possumus;* p. 222) They evoke equally voluntary, involuntary-conscious and unconscious movements, among them—that of the heart!

> In the heart one can observe the third movement of the spirits. It depends on the influx of the animal spirits, but is not in our power ... it is called "natural," not because it depends on anything but the animal spirits, but because nature cannot maintain itself without it.[106]

Therefore, an interruption in the influx of spirits in the cardiac nerves must stop the movement of the heart (*motus aboliti in corde causam, esse spirituum interceptione in nervis;* p. 137); if excised hearts nevertheless continue to beat for a time, this is to be explained by residues of spirits in the fibers of the heart, which stretch and shorten them (ibid.)

Thus, Highmore accepts the Cartesian reinterpretation of "natural" movements (while he abruptly invents a new reason for this appellation!) and gives for cardiac activity in effect the explanation that we have met first in Regius. As we noticed earlier, since the forties, Regius had been an important mediator for the spread of Cartesianism in England; it is to be assumed that his views were also known to Highmore. But even if that were not the case, the agreement between the two views would still be an expression of the mechanistic thought-style that was becoming prevalent everywhere.

2. *Thomas Willis (1621-75).* Thomas Willis, Professor of Medicine in Oxford and later co-founder of the Royal Society in London, belongs, along with Franciscus Sylvius, among the chief proponents of iatrochemistry in the seventeenth century. Strongly influenced by Robert Boyle and the French mechanists, he developed a unique combination of chemistry and atomism, the influence of which reached far beyond England. At the same time, his basic philosophical position bears clearly Cartesian features.[107]

In his extensive writings, Willis's views underwent some changes, which, however, we cannot follow here. Since Willis's later works were heavily influenced by his students Lower and Mayow, we shall confine ourselves here to the earlier writings, the *Diatribe duae medico-philosophicae* (*De Fermentatione* and *De Febribus,* 1659) and the *Cerebri Anatome* (1664).

At the beginning of the *De Fermentatione,* Willis explains his sympathy with atomism and its rejection of occult qualities, sympathies, and other *"ignorantiae asyla"*.[108] Nevertheless, the five basic elements he recognizes still retain a qualitative-chemical character. It is spiritus, sulphur, sal, aqua, and terra from the combination of which all bodies and their alterations are to be derived (p. 4). The spirits represent the finest ether-like substance and, at the same time, the basic stuff of vital processes, insofar, namely, as they contribute essentially to organic fermentations (p. 5). By this, Willis understands "an inner movement of the particles, that is, of the principles (!) of each body, with the tendency toward its perfection or its transformation."[108]

The central fermentation takes place in the interior of the heart, where the particulate configuration of the blood is loosened through increased heat so that the particles of spirit, salt, and sulfur become volatile and drive the blood like boiling water out of the heart into the vessels (p. 24).

This corresponds in essence to the Cartesian theory of the heart, which Willis supports even more explicitly in the *De Febribus*. Here, he first undertakes a basically chemical analysis of the blood (pp. 99 ff.). From the disintegration of the blood in vitro, he infers, unlike Harvey, a heterogeneity also in vivo:

> The blood ... contains heterogeneous particles of differing shape and energy, through the interaction of which in the mixed state the movement of fermentation is continously maintained.[110]

The description of the circulation proves to be comparatively simple: " ... as in a hydraulic machine, [the blood] is constantly driven in a circle due to the structure of the heart and vessels" (... *cordis, et vasorum structura, velut in machina hydraulica, constanti ritu circumgyretur;* p. 110.). The reaction of the blood in the heart, on the other hand, is, as Willis admits, harder "to explain mechanically"; still, Ent, Descartes, and Hoghelande had furnished explanations of it that were not improbable (p. 113). Consequently, the effervescence must result from a ferment in the cardiac chambers, which would set free the particles of spirit and sulfur so that they would break out with the blood into the arteries (*per aperto arteriarum ductus, qua data via, cum tumultu, et turgescentia ruunt,* ibid.) Willis attributes this change to a "nitro-sulfuric ferment", and thus combines the element of sulfur as bearer of all exothermic reactions (fermentations or combustions) with the saltpeter (*nitrum*) which, in the Oxford school of physiology, was becoming more and more the universal connecting link in breathing, heat, and combustion.[111]

In his next work, on the contrary, the *Cerebri Anatome*, we see Willis taking his distance from the Cartesian theory of the heart and pursuing the escape route already familiar to us.[112] In the footsteps of

Descartes, he first takes up in his work a reinterpretation of the two non-rational parts of the soul. The *anima vitalis* or *flammea* corresponds to the chemical heat in the blood, the *anima sensitiva* or *lucida* to the spirits in the nervous system (p. 71):

> ... it is the distribution and arrangement of the animal spirits that constitutes the essence of the sensitive soul, which is indeed nothing but a certain union of these spirits, which, adhering to and interlaced with one another, like atoms or fine particles, take on a certain form.[113]

Willis describes in detail the distillation and preparation of the spirits in the cerebrum and cerebellum, the former responsible for sensation and voluntary movement, the latter for unconscious functions like heart beat and respiration. Through this local differentiation, Willis succeeds for the first time in centralizing the functions of the body— the formerly "natural" functions are now only unconscious "animal" (spirit-governed) movements—but nevertheless, at the same time explaining the difference in their susceptibility to influence by the will.[114]

Further, Willis develops a chemical theory of muscular movement: the union of the animal spirits with sulfuric particles of blood issues in an "explosion" and allows the muscle to swell (p. 141f.). And he applies this theory to the heart as well. In diastole, the spirits in the cardiac parenchyma are united with the particles of blood streaming in; the additional entry of spirits through the vagus nerve then releases the explosion and thus systole:

> ... the spirits ... within the fibers and nervous filaments of the heart take up a great number of sulfuric particles from the blood that is streaming in; when they are then ... shaken and brought to the exploding point, so to speak, ... by those spirits (of the vagus nerve; author's note) ... there follows, when their walls are forcefully pushed inward, the systole of the whole heart, and the blood is thrown like a shot out of its chambers.[115]

Thus, Willis displaces the explosive fermentation from the chamber into the parenchyma, from which the blood is now driven out by systole. But thus with him, too, the heart loses its independence. It is a "machine," an "automatism" (p. 171) behind whose apparent *vis pulsifica* stands only the central regulation by the cerebellum (p. 169); it is a partial mechanism within the "thermal automatism," the great furnace of the body.[116]

Thus in the very alteration of his views, Willis exemplifies in a special way the development of physiological thought that we have been describing.

3. Richard Lower (1631-91). It was Thomas Willis's student Richard Lower who, in his *Tractatus de Corde* (1669), composed the most significant cardiological work since Harvey's *De Motu Cordis* and in it definitively disqualified heating and fermentation as the cause of cardiac movement, whether of systole or diastole.[117] His arguments are impressive: the complex fibromuscular structure of the heart on the basis of which it cannot be a mere container; its demonstrably only average temperature; the inert nature of the blood, the brevity and regularity of the sequence of heart beats, which make fermentation highly improbable; the questionable source of a fermenting material (systole leaves no space, but the walls of the chamber show themselves to be impenetrable); and the continued beating of the heart filled experimentally with other fluids or isolated—all this led Lower to the same conclusion:

> I am so far from believing that the movement of the blood depends on any kind of blazing up in the heart, that even its warmth does not seem to me to be in any way ascribable to the heart.[118]

In order to complete his argument, however, Lower also turns, on the other side, against his revered Harvey. Harvey's observation of a primary movement of the blood could in fact call into question the pumplike effect of cardiac contraction as the only cause of the circulation. Hence, Lower (like Highmore before him, cf. note D 101) reverses the sequence in the development of blood and heart:

> . . . the (cardiac) vesicle was there earlier than the blood . . . (it) is from the beginning not only the receptacle for the fluid changed into blood, but the machine that moves it.[119]

And, finally, as to the terminal movements of the blood in the quiescent right auricle that were mentioned by Harvey, Lower considers that these "arise not through any inner movement of the blood, but through the peristalsis of the vessels caused by the spirits, in the same way that the spirits in the muscles . . . continue that trembling movement even after death."[120] He continues:

> Since then . . . it is established, that the movement of the heart does not depend on the blood, we have next to say with what instruments and devices it is produced.[121]

These means are the fibers, the nerves, and the spirits flowing through them, which are mechanically extracted from the blood by the brain (p. 16). Lower turns against the expansion and explosion theory of muscular movement, but hesitates to specify the exact mechanism of cardiac contraction; in the last analysis, its explanation is reserved for God (*cum . . . dei solus . . . motum quoque eius cognoscere praerogativa sit*, p. 85). This much is certain however: the heart does not receive its force from within, but "from above, as it were from heaven" (*vim illam . . . superne et velut coelitus a capite ad illud descendere*, p. 85f.)—namely, from the animal spirits that arise in the cerebellum. For with the ligature of the vagus nerve, Lower observes, the heart falls into convulsions and finally abandons its activity (p. 86).

Thus, the movement of the heart is effected exclusively by the stream of spirits through the nerves (*motum cordis ab influxu in nervos . . . solum perfici*, p. 88) and is also regulated in the same way: emotions alter the pulse in this way; and vital spirits streaming in, in either too small or too great an amount or with too much force can have a deleterious effect on the movement of the heart (p. 124).

> Thus the movement of the blood and of the heart are wholly dependent on the brain . . . the brain rules over all parts of the lower body like a king over his subjects, and it steers and guides and governs everything through its will and command.[122]

With Lower, as we see, the heart loses altogether its special place in the organism and becomes a purely mechanical, dependent organ. In principle, Nicolaus Steno had already anticipated this when he included the heart in his anatomy of the musculature:

> But if it is certain . . . that the heart lacks nothing that the muscle has and that there is nothing in the heart that is lacking in the muscle, then the heart can no longer be a substance of a peculiar character, and consequently also nothing like the seat of the fire, of the inborn heat, of the soul, or the progenitor of a certain fluid like the blood, or the source of certain spirits . . . But it may be that the advocates of the heart will tell me that all this has

> to do only with the substance of the heart, not with its chambers, which contain the noblest part of life, namely, the blood. However, insofar as I can judge such matters, I fear that all arguments about the value of parts of the body are mere figures of speech.[123]

Lower carries this still further. He not only declares that the blood in general rather than the heart is the carrier of heat (*sanguini . . . debemus quod Cor ipsum caleat, quod corpora nostra calore suo actuet et vivificet;* p. 73)—this would still be compatible with Harvey; but for the first time, Lower also places the heating of the blood in the lung. He was led to this conclusion by the vivisections he had carried out with Hooke in 1664 and 1667, in which air was blown into the lung. These experiments showed that the change in color of the blood takes place, not in the heart, but during passage through the lung—according to Lower, through the reception of a *pabulum nitrosum*, a saltpeter-like fuel from the air.[124]

Basically, this meant the end of the conception of the central vital heat in the heart—even though it would occasionally be put forward again in isolated cases. Of importance in making this step possible was the Cartesian reinterpretation of innate heat as a purely physico-chemical process which was thus made available for experimental investigation. Only after the material-objective side of the vital heat had come to the fore was it possible to call the traditional conception into question on this level as well.

Although the heart is the central subject matter of Lower's work, it has now been robbed of its most significant function; nor is it seen in a reciprocal relation to the blood, but is degraded to an externally functioning, mechanical propulsive organ for the blood, one that is subordinated to the new ruler in the organism, that is, the brain.

4. John Mayow (1641-79). The Five Treatises of John Mayow, which appeared in Oxford in 1674, once more combined the English physiological tradition and the Cartesian doctrine of nature, the experimental and the deductive-speculative methods, in a bold design, which in a number of respects placed Mayow far ahead of his time.[125] Like Descartes, he is looking for a common, non-organic principle of chemical and physiological processes and finds it in the "niter of the air"—a "pre-idea" of oxygen:

> That the air surrounding us, which because of its fineness eludes the glance of our eyes, and appears

> empty when we scan its space, is filled with a salt, as it were universal, of the nature of saltpeter, that is, with a vital, fire-like and highly fermentative spirit—this will become clear in what follows.[126]

The air is for Mayow no longer a homogeneous, inert element with primarily qualitative effects ("cooling,", "animating"), but a mixture of corpuscles with the "particles of saltpeter in the air" (nitro-aereal particles or nitro-aereal spirits) as their most important component. To these particles, Mayow ascribes combustion, fermentation, respiration, vital heat, and the activity of heart and muscles. On the foundation of Cartesian physics—he borrows Descartes's laws of nature (S p. 887f., 172; M p. 104), the communication of motion through pressure and impact alone (S p. 86f.), the ether as first mover (S p. 87ff.) and essential elements of the cosmology of vortices (S p. 128-134)—Mayow first describes exothermic reactions and fermentations as the interaction of saltpeter and sulfur particles, followed by the liberation of the ether from the interstices of coarser matter (S p. 172f., 176f.). In this, he is following Descartes as well as in the fact that he equates combustion (and thus fire) and fermentation on principle as inorganic, heat-producing reactions (*ignis . . . nihil aliud esse videtur quam particularum nitro-aerearum, et sulphurearum quam maximam fermentationem,* S p. 176, cf. also p. 146f.).

Now, in the saltpeter particles in the air, we have the long sought common demoninator of combustion and life (*ignis et vita iisdem particulis aeris sustinetur,* p. 108). For they are united with the blood in the long and release an arterial fermention. There is no longer any cooling function of the atmospheric air; on the contrary, the air itself now feeds the fire of life. After the temporary separation of air and vital heat in Harvey, the hyphen in "nitro-aerial spirit" thus signifies the renewed unification of the "Heraclitean" and "Anaximenian" traditions in the manner of Galen—but now in the dress of atomistic chemistry, which sees the same principle at work in air and fire and conceives of the organism on the model of the non-organic.

However, that does not exhaust the task of atmospheric saltpeter; Mayow also makes it the "primary instrument" of living movement (S p. 47). Muscular contraction is based on a torsion (*contortio*) and shortening of the fibers, caused by the heat of reaction of saltpeter and sulfur particles (M p. 78ff.). But according to Mayow it would scarcely be possible to localize both agencies in the blood (R p. 304); in order for the reaction to take place precisely in the muscles, the saltpeter would have to enter on a separate path via the nerves. The animal spirits, Mayow infers conclusively, are thus nothing but the nitro-aereal spirit, which is extracted from the blood in the brain and is conducted to the muscles, including the heart (M p. 24). There, its reaction with the particles of sulfur in the blood evokes muscular movement. Life therefore consists, according to Mayow, in the central distribution of these spirits through the nerves:

> ... in order to complete this, the heart beat and the streaming of the blood to the brain is needed; and breathing seems to lead especially to the movement of the heart. For it is probable, that this atmospheric salt is needed for every muscular movement; so that the heart beat cannot take place without it.[127]

Every muscular action needs particles of saltpeter; respiration and heart beat serve to supply it, and their acceleration in the case of bodily exertion thus finds its explanation (M p. 24f.) Mayow has simply followed Descartes's argument through to its logical conclusion. If the movement of the living, which, from a physicalist perspective, looks at first like an "obscure movement," (M p. 4), can be traced back to non-organic movement, then in the last analysis, its source of energy must lie outside the organism. The heart can no longer be the autonomous principle of motion of the organism. It now has only the mechanical function of ejecting the blood (*illusque munus in sola contractione sanguinisque expulsione consistere,* R p. 303).[128]

It is interesting to ascertain that, for Mayow, this cancellation of vital autonomy is by no means to be taken for granted:

> Against what we have said it might be objected that the animal spirits represent the most excellent part of the body; wherefore they cannot come from the air, which is something external, foreign to the body, but rather they must be compounded from the finer particles of the blood ... And it would not be fitting for the so marvellous artifact of the living machine to be set in motion by an external principle.[129]

Harvey had spoken in the same sense against the derivation of the spirits from the air and had pleaded for the inherence of the vital principle in the blood. Mayow defends himself by pointing out that the immense losses of spirits could only be replaced from outside; and that the ingenuity of the body is evident precisely in the fact that the ordinary air can produce in it such amazing effects (ibid.). Thus, the universal inorganic substrate, as Descartes had conceived of it, now triumphs over the special "stuff of life."

In this way, the foundation has been indicated on which a century later oxygen will become the principle of heat and of life.

C. OTHER WRITERS: THE LATENCY OF THE VITALISTIC ASPECT

We have been considering the influence of the Cartesian thought-style on Dutch and English physiology; however, it was by no means limited to those countries. The mechanistic conception of the movement of the heart and blood, especially the association of cardiac action with the central distribution of the spirits, was dominant in European physiology for about a century. Nevertheless, in addition, we will now discover in a series of writers (chiefly in other countries), partly in open contradiction to their mechanistic starting point, elements of a "vitalistic" explanation of the circulation. It seems that precisely the riddle of the "spontaneity of the heart" kept giving rise to such responses. Here, central Harveyan concepts also reoccur in partly mechanistic contexts—a phenomenon that was characterized earlier as indicative of the "latency" of a point of view. The consideration of these authors must here be kept to the barest necessity.

Jan de Wale (Johannes Walaeus, 1609-49), who studied and taught in Leyden without belonging to the Cartesian school, in his "Two Letters on the Movement of the Chyle and the Blood" (1641), supported Harvey's discovery with his own experiments and demonstrations. In this work, he pleads for a mechanistic point of view in physiology; all bodily functions are on principle to be explained mechanically since, in accord with reason and the senses, we have here to start with "machines."[130] His exposition of the circulation of the blood corresponds to this point of view as far as the rejection of a spontaneous movement of the blood is concerned. Like every inanimate object, the blood can have no spontaneous movement through an "ingrafted virtue" (*nec in ulla alia re inanimata talem spontaneum motum licet videre;* p. 553). The motion observed by Harvey must therefore be attributed to contractions of the *vena cava* or of the auricle. For the heart, on the contrary, Wale does still admit a "natural movement," by which it responds to irritation. The expansion of the blood heated in the heart constitutes such an irritation; the heart tries to remove this "disturbance" and contracts (*[calore] mutatus sanguis cordi . . . sit molestus;* p. 554 and 557). This is in accord with Harvey's explanation; on the other hand, invoking Galen, Wale also reintroduces local "attraction" as a supplementary cause of the circulation. Only it could assure the streaming into the tissues of the blood, which nourishes them and provides them as appropriate with certain of its components (pp. 555, 562).

In a somewhat different form, attraction also plays a role in the case of Hermann Conring (1605-81), Harvey's first follower in Germany (*De sanguinis generatione et motu naturali,* 1643).[131] Conring considers the pressure exerted by the heart insufficient to explain the whole movement of the blood, particularly since in arterial ligatures the corresponding venous blood nevertheless moves toward the heart—even after the heart's excision! (p. 421). The chyle, too, moves without any impulsion through the lymphatic vessels. Conring therefore postulates an attractive effect of the innate heat in the heart. "The blood tends in this movement to go from colder to warmer places," from the periphery to the center (*sanguis videlicet hoc motu tendit a minus calidis ad magis calida,* p. 424). Thus, Harvey's centripetal tendency of the blood reemerges here as well as the necessity of its conquest by the heart (*cum calor attrahat potius sanguinem quam distribuat, ex eodem foco vi quaedam propellendus fuit sanguis;* p. 430).

The Danish anatomist Thomas Bartholinus (1616-1680), in his widely distributed *Anatomia reformata,* also provides such "unmechanistic" interpretations of the movement of the heart and blood.[132] More plainly than Wale, he describes the cause of cardiac activity as a double one, composed of the motion of the blood and the pulsative faculty of the heart (p. 252). The blood swells up and "irritates" the heart into contracting (p. 257). "Thus the heart is constantly moved by the blood, like a mill wheel that is driven by the constant impact of the water, and on the other hand, stands still without it." Therefore, if the flow of blood to the heart is cut off in an experiment, its movement breaks down and becomes irregular.[133]

But the heart must also be credited with its own pulsative faculty, "so that it supports and directs the flow of blood in and out, which would otherwise take place on its own, without regulation . . . the pulse would always be irregular, as would the influx of blood, if it were not governed by a faculty."[134] Thus, the heart is not only a strengthener of impulses; it also provides rhythm for the flow of blood. Bartholinus rejects Regius's and Hoghelande's mechanistic substitution of the pulsative faculty by the animal spirits or blood particles in the parenchyma—not only on account of the movements of the excised heart, but also (as far as the animal spirits are concerned) because the heartbeat is a "natural," involuntary movement and not an animal movement.[135] Thus, Bartholinus holds firmly to the traditional distinction and, very like Harvey, interprets cardiac action as an autonomous reciprocal relation, not centrally governed.

Even in England, despite the dominant mechanistic influence, an after-effect of Harvey's vitalistic conception of the movement of the heart is to be found in two places. On the one hand, surprisingly enough, it is to be found in Walter Charleton's *Exercitationes Physico-Anatomicae de Oeconomia Animali . . . mechanice explicata* (1659). Although he was one of the chief proponents of atomism and mechanism in England, Charleton adopts Bartholinus's interpretation of the movement of the heart, even borrowing his language, and comes to the conclusion:

> The blood moves the heart through its mass; and the heart drives the blood out by its contraction; thus the movement of both is perpetuated . . . The pulse originates from a double agency, namely, through the blood and the heart . . . The heartbeat arises from the blood, which irritates the heart and so makes it contract.[136]

On the other hand, it is Harvey's student Francis Glisson (1597-1677) who embeds the interpretation of cardiac activity as a living polarity within the framework of the first philosophically grounded physiology of irritability. In his *De Ventriculo et Intestinis* (1677), Glisson, who had at first spoken only of an irritability dependent on the nerves, develops, under the influence of Harvey's *De Generatione*, the theory of a primary "sense" of the tissues, a "natural perception" in contrast to conscious "sensitive perception."[137] Natural perception is the expression of the elementary, pre-mental vitality of living tissues, which is merely built upon and modified by the higher, central soul. Glisson associates this with Harvey's "plastic virtue":

> Natural perception . . . knows an almost infinite amount that remains hidden to sensory perception; and it knows the whole order of the body— which it has itself formed—, the functions of the parts and the way everything is to happen.[138]

According to Glisson, every single fiber possesses a directed sensitivity ("perception" and "appetite") for positive and negative irritants; from this and from elementary motility (the "motive faculty"), the irritability of the fibers results (pp. 168ff.). The higher functions of consciousness, too, arise on this foundation. The nerves can release the activity of the voluntary muscles only through the fact that they perceive the movements of the brain through natural perception (p.

185). Nevertheless, natural perception becomes really evident only in unconscious functions like the activity of the heart:

> The heart beat neither occurs on the ground of sensory perception nor is changed by such perception. The fibers of the heart are irritated by turns through the pulsing vital blood in its chambers, are stimulated to contract and thus produce the pulse; then once more the irritation abates, the fibers relax and return to their natural position. It cannot be denied that what occurs here is plainly an irritation of the fibers. For the rhythm of the pulse varies correspondingly, as is shown in the differences of the pulse in fever and other illnesses ...[139]

Thus, for Glisson, the activity of the heart becomes the favorite example of an autonomous vitality whose polar structure (and not some central regulation) both makes possible the rhythm of the pulse and conditions its variability. It is precisely through Glisson that Harvey's ideas continue to have influence in the eighteenth century.

To conclude this overview, let me refer again to two authors with a basically mechanistic point of view, who, however, on the question of cardiac activity, mix in vitalistic elements: Giovanni Borelli in Italy and Johannes Bohn in Germany.

In his work *De Motu Animalium* (1680-81), Giovanni Borelli (1608-79), who is allied in his thought to Cartesianism, but belongs primarily in the tradition of Galileo and Santorio, considers all the outer and inner movements of the organism under mathematical and physical principles in accord with the Cartesian concept of a law of nature. Nature, he writes at the beginning of his treatise, can do nothing that goes beyond the divinely prescribed, necessary laws of mechanics.[140] Thus, muscle movement too is a "mechanical operation" (p. 41) and works, rather as it does in Willis, through the explosive, expanding reaction of the nervous fluid (or animal spirits) with a *succus tartareus* from the blood (p. 46f.):

> The mechanical process by which such a heating and effervescence is effected does not differ from those that take place in ordinary fermentation. In all these processes motive forces are not created de novo, but forces that have previously been restrained are set free, so that they can exercise their natural function.[141]

The heart, too, works in this way, not on the ground of an "incorporeal faculty" (p. 112), but as a mechanical press or pump driven from outside by the blood and nervous fluid (*instar praeli/ veluti ab embolo syringae,* p. 66f.). Borelli also calculated the physical time of cardiac activity, as well as peripheral resistance (p. 108) and gave an exact measurement of cardiac temperature (p. 137f.).

Thus, as in the Cartesian tradition—with which Borelli is thoroughly familiar—cardiac activity seems to be assimilated to the movement of the muscles regulated by the central nervous system.[142] What remains puzzling, however, Borelli continues, is the special autonomy and involuntary character of cardiac activity (p. 110); what causes the constant recurrence of this movement? He examines and rejects various alternative mechanisms like the pendulum or the valves in the nerves postulated by Descartes and finally arrives at the notion that the nervous fluid leaves the very narrow cardiac nerves only drop by drop.[143] Thus, it would indeed be possible to explain the movement of the heart as that of an automaton (*motum cordis fieri posse . . . necessitate organica, non secus, ac automata movetur;* p. 114); but Borelli confesses that he is not yet satisfied. For since differing psychological states have unambiguous effects on the activity of the heart, and since now one, now another cause cannot be assumed for the same function, it must nevertheless be the "sensitive faculty of the soul itself" that is responsible for the action of the heart.[144]

Surprisingly, Borelli interprets this by analogy to the learning of voluntary movements, which transpire "in flesh and blood": for example, movements to keep one's balance, or in playing an instrument and the like. Cardiac movement, too, is "learned" by the soul: at the beginning, in the embryo, which exists in preformation, the sensitive faculty perceives through the nerves the swelling blood in the heart (*percipiat, mediantibus nervis, gravedinem . . . in corde, a sanguineo succo ibidem turgente;* p. 116); it tries "to eliminate this disturbing content." From this constantly repeated reaction, there finally arises a habit:

> Thus a person would not seem inept, let alone ridiculous, who wondered whether the pulse of the cardiac muscle could follow, not from blind mechanical necessity, but through that very power of the soul through which all the muscles of the limbs are moved according to arbitrary precepts, but which can become unconscious habits.[145]

This unusual interpretation of cardiac activity—on the basis of a condition of stimulation that was perceived by the soul and had become habitual—shows, expecially in contrast to Glisson, to what an extent the Cartesian-dualistic thought-style had altered the boundary conditions of physiology. Borelli does indeed see his purely mathematical-physical principles of explanation as inadequate at this point; still, he is unable to recognize even in the embryonic heart any autonomous vitality and sensibility. As a second principle, there remains, in addition to the material world, only the central world of consciousness or soul, which, despite its "powers," is considered unitary, and for which now every movement must be, at least originally, "willed." Yet, on the other hand, this conscious world remains tied in its activity to brain and nerves—a condition which can be fulfilled through preformatist embryology.

As early as Sylvius, we have seen how close the way that the Cartesians tied "natural" movement to the central distribution of the spirits came to the notion that these movements might have something to do with the "will," that is, with the rational soul. Borelli carries this idea further—to a consequence that would no longer be in the spirit of Descartes: the cardiac nerves, in the first instance only the material substrate of the feelings, then the mechanical cause of cardiac movement itself, now even become the instrument with which the soul is able to control the vital function of the heart. But this already breaks through the limits of the Cartesian res cogitans, for which separation from the automatisms of the body is an essential criterion.

It is, then, the mechanist Borelli who, in a certain sense, anticipates the animism of Georg Ernst Stahl (1659-1734). Stahl, however, dispenses with the mediation of the nerves and sees himself therefore confronted even more radically with the problem of how the soul, which is opposed in unified form to the body, is now in a position to control immediately each of the body's functions.

Johannes Bohn (1640-1718), Professor of Anatomy and Medicine in Leipzig and probably the most significant German physiologist of his time, again illustrates the predominance of the Cartesian point of view in theoretical medicine at the end of the seventeenth century. With many references to Descartes, Bohn makes Cartesian dualism the explicit foundation for his *Circulus anatomico-physiologicus* (1686).[146] The body is a machine whose organs resemble wheels, levers, springs, and ropes, and which has been assembled by the divine "mechanic" (p. 45, p. 47); there is no substance other than body and mind (p. 50).

"*Omne corpus quod movetur, ab alio movetur.*" (Every body that moves is moved by another.) Bohn quotes this scholastic principle, but rejects the soul as a moving principle: it is not responsible for the economy of the body and modifies the running of the machine only occasionally (p. 45, p. 50). Instead, the actions of the body are determined independently of the soul through the arrangement of the organs, especially the brain, as well as through irritations from without, so that the different movements, according to Descartes, originate indeed in man, but not through him."[147] Just like the beast, therefore, he needs a "material and mechanical principle of motion," which Bohn localizes in the heart and brain:

> What the pump or main piston accomplishes in a hydraulic machine, with the circulation of an ambient fluid, the heart accomplishes in the living machine: just as the former provides the first impulse for the water, so does the heart for the blood, by sending it out and propelling it; but if both (machines) become exhausted, their fluid stagnates.[148]

The mechanical view of the heart persists in the interpretation of the circulation (*sanguis . . . omnem sui tendentiam a sola cordis impulsu . . . recipit,* p. 324; likewise on p. 106). The heart possesses no special "vitality": it is neither the source of vital heat nor the seat of the sensitive soul (p. 112). The passions, which are allegedly kindled and perceived there, are in truth more violent movements in the brain, which are projected onto the heart via the animal spirits as Descartes has explained (p. 113). The sole function of the heart is movement (*quicquid cor agit, motu agit,* p. 99).

Nor can this movement be explained through a pulsative or vital faculty (p. 103). From the apparent autonomy of the heart, one must not, like Borelli, infer any other cause of movement than that which obtains for the muscles:

> . . . it is rather to be assumed that it [the heart] is moved with same mechanical or organic necessity, that is, through the adaptation and expansion of the fibers, consequently also through the animal spirits.[149]

This is demonstrated also by the effects of the ligature of the vagus in the experiments of Willis and Lower. Yet, in what follows, Bohn explains the rhythmic sequence of cardiac action through a principle for which there is really no basis in his physiology: it is the "stimulus" of the blood that keeps pushing in, which the cardiac fibers "perceive" and respond to with contractions (*stimulum . . . persentiscunt,* p.

105). This stimulus does indeed act only as a strengthening factor; yet its significance is shown by the fact that an interruption of the blood supply permits the pulse to slacken (loc. cit.).

In the section on muscular movement, Bohn investigates this problematic once more in a more general form (pp. 448ff.). The distinction between "natural" and "animal" movements, he says, is really not appropriate, since both depend equally on the animal spirits, or on the muscles as the spirits' passive instruments (*instrumentum huius motus organicum seu passivum,* p. 451). But then how is the automatism of "natural" movements like that of the heart to be explained? Bohn first discusses the solutions of Willis (differential localization in the cerebrum and cerebellum) and Borelli (habituation to an originally "voluntary" movement responding to a stimulus). He is inclined, in fact, toward Borelli's explanation—the cause of natural movements is then a constant inner state of stimulation (p. 467)—but is nevertheless reluctant to ascribe the perception of these stimuli to the rational soul. Thus, in the end, Bohn comes to a quasi-vitalistic conception, which is scarcely different from Glisson's:

> . . . every living fiber has a kind of instinct, through which it is stimulated to contract or to move: this sensing or perception of the stimulus we must see not as a kind of touch, but as its analogue and as the determinant of movement . . . That the heart too is not without its stimulation, and consequently its power to feel, we can learn from what was said above.[150]

In this way, two causes of the movement of the heart are opposed to one another: the central distribution of the spirits on the one hand and, on the other, the stimulation of the cardiac fibers by the blood; and in the last analysis, their relation to one another remains unexplained. In Bohn's case too, as we can see, the peculiarity of the heart breaks through the prescribed framework of dualistic physiology. Bohn lacks the concepts and principles needed to grasp this peculiar activity adequately at its own level, concepts and principles that the Cartesian point of view had excluded from the study of the organism.

NOTES TO PART D

1. Through his friend, the influential Father Marin Mersenne (1588-1648), Descartes learned of Harvey's theory soon after the appearance of

the *De Motu Cordis*, but he read the book only in 1632, after the composition of his own physiological essay, the *Treatise on Man* (*Traité de l'Homme*) (1632). This manuscript, together with the later *Description du corps humain* (*Description of the Human Body*, 1648), was published only posthumously under the title *De Homine* in 1664. The text is given in *Oeuvres*, ed. C. Adam and P. Tannery, vol. 11, pp. 119-202 and 219-290. The two works are referred to as TH and DCH respectively. Where passages cited are included in the English translation by J. Cottingham, R. Stoothoff and D. Murdoch, Cambridge: Cambridge University Press, 1985, this version has been followed; it is referred to as CSMK.

2. *Principia Philosophiae* (1644), I, 30, AT VIII-1 16-17, CSMK I 203.
3. *Meditationes de Prima Philosophia* (1641), AT VII 34, CSMK II 22.
4. *La Dioptrique*, AT VI 84ff., CSMK I 153f.
5. *Les passions de l'ame* (1649), I, 23, AT XI 346, CSMK I 337. The comparison of the soul with a blind man reads as follows: "when our blind man touches bodies with his stick, they certainly do not transmit anything to him except in-so-far as they cause his stick to move in different ways . . . thus likewise setting in motion the nerves in his hand, and then the regions of his brain where these nerves originate. This is what occasions his soul to have sensory perception of just as many different qualities in these bodies as there are differences in the movements caused by them in his brain" (AT VI 1214, CSMK I 114). Cf. Med. VI, AT VII 83, PP IV, 198, AT VIII-1 321-1, CSMK I 284-5.

The Cartesian reinterpretation of perception as the encoded neuronal transfer of information to a cerebral consciousness that in itself is "blind" to the external world has been wholly assimilated to the modern point of view: according to this view, the most divergent sense qualities are all based on the same electro-chemical processes in receptors and neurons. Encoding occurs uniformly in the shape of transmitted action potentials. And what is encoded is not diverse qualities of the external world (such as 'red,' 'blue,' etc.), but again only different configurations of a world of clouds of atoms, waves, and fields of energy.

6. In the *Principles*, we read that a man who is a philosopher "should put his trust less in that senses, that is, on the ill-considered judgment of his childhood, than on his mature reason" (I, 76, AT VIII-i, 39; CSMK I 222, with revision). Descartes applies this principle in many places in his work on natural philosophy, for example, in the account of the laws of nature: "These matters do not need proof since they are self-evident; (the demonstrations are so certain that even if our experience seemed to show us the opposite, we should still be obliged to have more faith our reason than in our senses)" (AT VIII-1 70, CSMK I 245; bracketed portion in French translation only, AT IX-2 93). Similarly, in Descartes's account of the activity of the heart: experiment "seems to make it quite certain" that, contrary to his view, the heart contracts when active; "[n]evertheless, all that this proves is that observations may often lead us astray when we do not examine their possible causes with sufficient care" (DCH, AT XI 242, CSMK I 317).

7. D.M. Clarke, *Descartes's Philosophy of Science* (Manchester: Manchester University Press, 1982), esp. pp. 63-70.

8. Despite the basically conceptual construction of his physical world of particles, Descartes expected confirmation of his views through the future progress of microscopy; cf. *La Dioptrique*, AT VI 226f. It should also be noted that, among other things, the infinite divisibility of matter distinguishes Descartes's position from that of contemporary atomism, with Pierre Gassendi as one of its important advocates (Cf. PP IV, 202, AT VIII-1 325, CSMK I 287-8).

9. This, too, in contrast to Harvey. On the "branchings" in deductive procedure, see the *Discours de la méthode*, AT VI 64-5, CSMK 144. On Cartesian science as based on models, see Disc. V, AT VI 41-2, CSMK I 132, and finally TH, AT XI 120, CSMK I 99: "I suppose the body to be nothing but a statue or machine made of earth, which God forms with the explicit intention of making it as much as possible like us."

10. Or as F. Alquié, writes, " As the prereqisite for technical action, clear understanding by its very nature fails to be adapted to ontology" (Alquié, *Descartes*, Paris: Hatier, 1956, p. 141).

11. On this cf. E. Zilsel, "Die Entstehung des Begriffs des physikalischen Gesetzes, in *idem, Die sozialen Ursprünge der neuzeitlichen Wissenschaft*, Frankfurt, 1976, pp. 66-97; also H. Schimank, "Der Aspekt der Naturgesetzlichkeit im Wandel der Zeiten," in *Das Problem der Gesetzlichkeit*, ed. Jungius-Gesellschaft der Wissenschaften, Hamburg, 1949, pp. 139-186.

12. "Car Dieux a si merveilleusement etably ces loiz, qu'encore que nous supposions qu'il ne crée rien de plus de ce que j'ai dit, et même qu'il ne mette en cecy aucun ordre ni proportion, mais qu'il en compose un Cahos . . . : elle sont suffisant pour fair que les parties de ce Cahos se démèlent d'elles même, et se disposent en si bon ordre qu'elles auront la forme d'un Monde tres-parfait" (*Le Monde*, AT XI 34-5, CSMK I 91).

13. ". . . et la connaissance de cet ordre est la clef et le fondement de la plus haute et plus parfaite science, que les hommes puissent avoir, touchant les choses materielles; d'autant que par son moyen on pourrait connaitre *a priori* toutes les diverses formes et essences des corps terrestres, au lieu que, sans elle, il nous faut contenter de les devinir *a posteriori*" (Letter to Mersenne, May 10, 1632, AT I 250, CSMK III 38). *Le Monde* was written at this time, and the *Traité de l'homme* as part of it.

14. H. Jonas, *Organismus und Freiheit*, Göttingen, 1973, p.23.

15. Cf. *Le Monde*, AT XI 37, CSMK I 92: " . . . by 'nature' here I do not mean some goddess or any other sort of imaginary power. Rather, I am using this word to signify matter itself . . . " (. . . *par la Nature je n'entens point ici quelque Désse ou quelque autre sorte de puissance imaginaire, mais . . . je me sers de ce mot, pour signifier la Matière même . . .*)

16. Cf. *Meditationes*, Replies to Fifth Objections, AT VII, p. 356: ". . . mentem enim non ut animae partem, sed ut totam illam animam, quae cogitat, considero." Also in the letter to Regius, May, 1641, AT III 371: ". . . contra logicam est, animam concipere tanquam genus, cuius species sint *mens, vis vegetativa et vis motris animalium . . .* "

17. Compare this with the view of Aristotle: "Something can be moved in two senses, either in respect to something else or in itself." "[The soul] can also be moved accidentally by something external. The living creature may be pushed by force. But what moves by itself according to its essence does not need to be set into motion by another, except accidentally" (DA I 3, 406a4ff., 406b6ff.). The scholastic maxim "*omne quod movetur ab alio movetur*" is naturally not to be understood in the causal-mechanistic sense; it does in fact also include the possibility that a body can be moved by its soul (and this in turn by some end or good). Nevertheless, the external relation of ("moving") soul and ("moved") body is already presupposed in this principle, a relation which can then be reinterpreted in the material-mechanistic sense. On this, see V. Larkin, "St. Thomas Aquinas on the Movement of the Heart, " *J.Hist.Med.* 15 (1960), pp. 22-30.

18. From the Conversation with Burman: "Deus corpus nostrum fabricavit ut machinam, et voluit illud agere ut instrumentum universale, quod semper operaretur eodem modo juxta leges suas." [CSMK translates "its own laws"; the German version, which seems more likely, makes it "his laws," that is, God's.] See also *Discourse* 5, AT VI 45-6, CSMK I 114.

19. In a certain sense, this very externality means precisely the *absolute dominion* of the end over its means: this too had been anticipated in the Galenic system, which made the body, in its functioning, wholly dependent on the soul. On the paradigm of the automaton in Galen and Descartes, see O. Temkin, "Metaphors of Human Biology" in R. C. Stauffer, ed., *Science and Civilization*, Madison: University of Wisconsin Press, 1949, pp. 167-194, esp. pp. 178ff.

20. On the same problematic, see also the letter to Regius, January 1642, AT I 502.

21. In view of current cybernetic or computer-based models of the body, his radical leveling of the difference between the natural and the artificial does not allow us simply to discard Descartes's mechanistic physiology today as obsolete. This is true not only because thoroughly cybernetic conceptions are found in Descartes himself (see below), but above all, because we think of the organism first and last as a "more complex" variant of the best automaton that we happen to be able to construct—and not as a phenomenon belonging to its own category. "Thus with the new model of mechanism, organisms can be termed mechanisms not because organisms lack self-relation, as Descartes argued, but because mechanisms have it. For... self-relation, neomechanism substitutes feedback" (M. Grene, "Life, Disease and Death: A metaphysical viewpoint," in S. Spicker, ed., *Essays in Honor of Hans Jonas*, Dordrecht: Reidel, 1978, pp. 233-263). "Animal and machine: each of the systems then becomes a model for the other... Organs, cells, and molecules are then united by a network of communication... The flexibility of behavior rests on loops of retroaction... Thus it is on the structure of a large molecule that the order of a living things is based. For reasons of stability, the organization of a chromosome comes to resemble that of a crystal... Heredity functions like the memory of a calculator" (F. Jacob, *La logique du vivant*, Paris: Gallimard, 1971, pp. 273-4).

Finally, let me quote Sir John Eccles, who, along with Sir Karl Popper, moves entirely along the tracks of Cartesian dualism: "This concept (of neuronal machinery) is purposefully chosen, in order to express the scientific assumption that the brain functions like a machine ... According to the philosophy represented in this book ... the brain is a machine of almost limitless complexity and subtlety." (K. Popper and J. Eccles, *Das Ich und sein Gehirn*, Munich, 1982, p. 282; passage occuring only in German edition.)

22. The most fundamental treatment of this topic is now that of A. Bitbol-Hespériès. *Le Principe de la Vie chez Descartes*, Paris, 1990. Still worth reading is A. Georges-Berthier, "Le méchanisme cartésien et la physiologie au XVIIe siècle" *Isis* 2 (1914), pp. 37-89; 3 (1920-21), pp. 21-58. Further: P. Mesnard, "L'esprit de la physiologie cartésienne," *Archives de Philosophie* 13 (1937), pp. 181-220; T.S. Hall, "Descartes's Physiological Method: Position, Principles, Examples," *J. Hist. Biol.* I (1970), pp. 53-79; G. Lindeboom, *Descartes and Medicine*, Amsterdam, 1978; also, the extensive notes in the edition of the *Treatise on Man* by K.E.Rothschuh and T.S. Hall, Cambridge, MA: Harvard University Press, 1972.

23. Cf. DCH, AT XI 244, CSMK I 318, Disc. 5, AT VI 46, CSMK I 134, as well as the letter to Newcastle, April, 1645 (?), AT IV 189.

24. Letter to Beverwijck, July 5, 1643, AT IV 5. Basically, as his contemporaries also noticed, Descartes is thus taking up again the Aristotelian explanation (De Resp. 20, 479b27ff). Compare also Georges-Berthier, op. cit., pp. 49-60.

25. "And is it not easy to understand the action that converts the juice of this food into blood, if we consider that the blood passing in and out of the heart is distilled perhaps more than one or two hundred times each day?" (Disc. 5, AT VI 53-4, CSMK I 138). Cf. DCH, AT XI 236.

26. In ER 164f., W. 139f. Harvey turns explicitly against Descartes's theory of the heart, and in particular, against its reversal of systole and diastole and against the uniformity of the principle of movement. On this debate, see Gilson, pp. 80ff., as well as R. Toellner, "The Controversy between Descartes and Harvey Regarding the Nature of Cardiac Motion," in A. Debus, ed., *Science, Medicine and Society in the Renaissance*, New York, 1972, vol. 2, pp. 73-89. Toellner emphasizes the different interpretations of the AT on the ground of the opposing points of view of Harvey and Descartes.

27. On this exchange, see A. Bitbol-Hespériès, 1990, pp. 83-90.

28. Plemp to Descartes, January 1638, AT I 497. The correspondence was later published by Plemp in his *Fundamenta Medicinae* (2d ed. 1644) as well as by the Dutch physician Johan van Beverwijck (1594-1647) in the *Epistolicae Questiones cum Doctorum Responsis* (1644).

29. Descartes to Plemp, Feb. 15, 1638, AT I 522f.; Plemp to Descartes, March 1638, AT II 53.

30. "... si haec instrumenta interdum ad hoc sola sufficiant, cur non semper? Et cur potius imaginaris illa in virtute animae agere, cum ipsa abest, quam ista animae virtute non indigere, ne quidem cum adest?" (Descartes to Plemp, March 23, 1638, AT II 65, CSMK III 94, modified).

31. Descartes's description of the circulation in the correspondence with Plemp also corresponds entirely to the conception just referred to of vortical motion as the successive displacement of particles: "Let us assume, for example, that BCF is an artery filled with blood ... (Descartes is referring to a diagram here—auth.) into which new blood is now entering from the heart A. It is easy to understand that this new blood cannot fill the space B in the opening of the artery without another part of the blood's receding in just this space B toward C, from there pushing other parts of the blood toward D, and so on to E so that in the same moment in which the blood rises from A to B, the artery must beat against E." (*Ponamus exempli causa BCF esse arteriam sanguine plenam ... in quam nunc ex corde A nonnihil novi sanguini ingreditur. Sic enim facile intelligimus hunc novum sanguinem non posse implere spatium B, quod est in orificio huius arteriae, quin alia pars sanguinis, quae prius implebat hoc idem spatium B, recedat versus C, indeque alias partes sanguine trudat versus D, et sic consequenter usque ad E; adeo ut eodem ipso instanti, quo sanguis ascendit ab A ad B, debeat arteria pulsare ad E.*) Finally, a similar description is also found in the DCH, AT XI 254.

32. DCH, AT XI 236. See also, the letter to Regius, November 1641, AT III 440ff.

33. "I do not deny life to animals, since I regard it as consisting simply in the heat of the heart" (*... vitam enim nulli animali abnego, utpote quam in solo cordis calore consistere statuo*). Letter to More, Feb. 5, 1649, AT V 278, CSMK III 366. See also PA I 5 and 8, AT XI 330, 333, CSMK I 329, 331.

34. On the following, cf. Bitbol-Hespériès, pp. 191-207.

35. This reduction can be followed in early manuscripts, in which at first there is talk of different spirits. On this, see H. Dreyfus-Le Foyer, "Le conceptions médicales de Descartes." *Revue de Métaphysique et de Morale* 44 (1937), pp. 237-286, esp. p. 243f.

36. Cf. Temkin (1951) as well as C. Galenus, *De Motu Musculorum* I 1, pp. 370 ff. in *Opera* (Kühn), vol. IV.

37. According to Descartes's letter to the Leyden professor A. Vorstius (1597-1663), the spirits belong to the third (terrestrial) element; they stand between air and fire (*omne corpus constans ex particulis terrestribus, materia subtili innatantibus, et magis agitatis quam quae aerem componunt, sed minus quam quae flammam, spiritus dici potest;* "every stable body made of terrestrial particles, bathed in subtle matter, and more agitated than the particles that make up the air, but less so than fire, can be called spirit;" AT III 687) Thus, they are not composed of ether (subtle matter), but are surrounded by it. As terrestrial particles, they are ultimately derived from nutriment (loc. cit.) Cf. DCH, AT XI p.248): "as the agitation of the first two elements supports that of the humors and the *spirits*." Nevertheless, the identification of Cartesian spirits with the first element or ether is an error that was later frequent and still occurs today (e.g. in W. Böhm, "John Mayow and Descartes, p. 53, *Sudh. Arch.* 46 (1962), pp. 45-68).

38. Bitbol-Hespériès has discovered in the *Theatrum anatomicum* of Caspar Bauhin is the probable source for Descartes's depiction of the pineal gland as "Gland H" (Bitbol-Hespériès, pp. 213 ff.).

39. Again, it is not for nothing that Descartes, in the *Meditations*, calls his body, face, hands and arms "this whole machine of limbs that can be seen in a corpse" *me habere vultum, manus, brachia totamque hanc membrorum machinam, qualis etiam in cadavere cernitur, et quam corporis nomine designabam*; (AT VII 26, CSMK II 17: "this whole mechanical structure of limbs . . . ").

40. Without, however, using the term "reflex." Ivan Pavlov above all has fully recognized the significance of Descartes for the development of neurophysiology and has referred to him in his reflex theory: "The physiologist must take his own path, where a trail has already been blazed for him. Three hundred years ago Descartes evolved the idea of the reflex. Starting from the assumption that animals behaved simply as machines, he regarded every activity of the organism as a necessary reaction to some external stimulus, the connection between the stimulus and the response being made through a definite nervous path; and this connection, he stated, was the fundamental purpose of the nervous structures in the animal body. This was the basis on which the study of the nervous system was firmly established" (Pavlov, 1960). In contrast to Pavlov, G. Canguilhem, in his careful analysis of the history of the reflex concept, has shown that it did not appear, in its strict sense, before Thomas Willis's *De Motu Musculari* (1670). However, Canguilhem grants to Descartes's mechanistic conception of movement that it " . . . prepares, *avant la lettre*, a context of comprehensibility for the reflex concept" (G. Canquilhem, *La formation du concept de reflexe aux XVIIe et XVIIIe siècles*, Paris, 1977, p. 169).

41. TH, AT XI 133. Galen distinguished between sensitive and motor nerves (cf. Hall 1972, p. 24); however, they were often identified with one another; the functional and anatomical differentation was first established by Bell (1811) and Magendie (1822). The idea of marrow fibers and nerve valves is genuinely Cartesian.

42. At this point, too, it is tempting to look back once more at Harvey's conception. For him, too, the motor nerves are also sensitive, but in quite a different sense: they are motor in so far as they mediate perception to the brain, that is, the perception of the *self-movement* of the muscles (*tanquam animal seperatum*); the brain is the choir master, and not the puppeteer or the controller of hydraulic pipes as in the automata of the fountains of Versailles (AT XI 130).

43. On this, compare L.J. Rather, "Old and New Views of Emotion and Bodily Changes: Wright and Harvey versus Descartes, James and Cannon." *Clio Med.* 1 (1965), pp. 1-15. Rather, too, takes the position that in Harvey heart and blood appear as the expressive organs of the emotions, while brain and nerves lost almost all significance in this context. "In this respect Harvey had few followers" (p. 6). Further development goes unequivocally in the Cartesian direction, that is, to an explanation of the psychological on the basis of humoral (later = hormonal) factors.

44. Galen and Vesalius considered them sensory, but others took them to be of a motor nature; cf. Hall (1972), p. 30, p. 70, and also Rothschuh (1969), p. 95.

45. Hammacher translates "aus dem Herz erhalte . . .", ". . . receives

from the heart"; but this does not correspond to the original ("que l'âme reçoit ses passions *dans* le coeur"), and misses precisely the point of the Cartesian argument: in the soul, not *in* the heart.

46. In this context, the emotional mechanism can vary with the individual so that the same object releases fear in one person and aggression in another (PA I, 39, AT XI 358-9, CSMK I 343). This phenomenon, too, is based on a differing *bodily disposition*, namely, as Descartes explains, an *imprinting* of the cerebral structure in early childhood (PA II 107-11, 136, AT XI 407-411, CSMK I 365-367, AT XI 428-9, CSMK I 375-376). The ethics of the *Discourse* rests essentially on the possibility of a consciously learned ("trained") novel combination of releasing object, movement of the spirits, and feeling brought about by the soul through voluntary opposing ideas (PA I 50, AT XI 368-370, CSMK I 348). In this way, Descartes also anticipates the modern theory of the conditioned reflex or learning.

47. Here, I agree with L.J. Rather: "Although there are immense differences in superstructure, both Cartesian and modern neurophysiological psychology rest on the doctrine of the reflex—providing a rationale for the machine-like functioning of the nervous system and the body that it controls—and some form or other of psychophysical dualism, which removes the mind or soul from its former position as a cause of bodily changes characteristic of the emotions and makes it into little more than a passive observer of these changes" (op. cit., p. 1). R. Carter also comes to similar conclusions: "It is not in the least attributing too much to Descartes to say that he envisaged a biochemical theory of moods and, to speak generally, what we today call 'inner states'" (R.C. Carter, *Descartes' Medical Philosophy: the Organic Solution ot the Mind-Body Problem*, Baltimore, 1983, p. 250).

48. " . . . the *Traité*, (*de l'Homme*—author) gives what William Harvey had failed to supply in the *De Motu Cordis*—an integrated physiological model on a par with that of the Galenic tradition . . . " (Sloan, p. 14f.). On the influence of the Cartesian models, see the previous note, as well as, for example, J. C. Kassler, "Man—A Musical Instrument: Models of Brain and Mental Functioning before the Computer," *Hist.Sci.* 22 (1984), pp. 59-92: "Descartes' concepts of neuro-muscular function—in their major outlines, at least—are still acceptable today . . . all we need to do is substitute nerve impulses for animal spirits, and synapse for pore" (p. 66).

49. Mendelsohn, who draws chiefly on in vitro experimental research, sees the reinterpretation of vital heat as starting with the Oxford physiologists (Boyle, Hooke, and Mayow) (op. cit., p. 62f.). However, the *conceptual principle* of this important reinterpretation had certainly already been anticipated in Descartes. If in the *Discours*, for reasons of custom, he attributes the original kindling of the "fire without light" in the heart to *God* (Disc. 5, AT VI 45-6, CSMK I 134), this attribution has only a provisional character—like that of the bodily machine itself, also "formed by God (see note D-8). In the same sentence, this fire is characterized as wholly inorganic; and in his embryology (a section in the "Description of the Human Body" of 1648), Descartes represents the melting of the two generative fluids as the *fermentational initial reaction* and basis of all further development

(DCH, AT XI 253-4). At first, the heart arises as a hollow form, insofar as the particles of the surrounding area are pushed outward by the reaction; then the vessels, the brain, and so on form as deposits. Thus this is, though in agreement on the course of epigenesis, the physicalistic counterpart to Harvey's ontogenesis from the vital autonomous pulsation of the first drop of blood:

> ... if one knew well what all the parts are of the seed of some particular species of animal, for example, of man, one could deduce from that alone, by entirely mathematical and certain arguments, the whole shape and conformation of each of its members; just as reciprocally, also, knowing some particulars of that conformation, one can deduce from that what its seed is (Disc. AT XI, 277).

Needham says: "... if anywhere ... we are to find the roots of physicochemical embryology, we must pause to recognize them here" (Needham, p. 156f.). Moreover, Carter has investigated in detail the striking agreements between Descartes's embryology and his cosmological vortex theory (op. cit., pp. 190-212); Descartes himself points out this parallelism (*Conversation with Burman*, AT V 170-171). These suggestions will have to suffice here; tempting though it would be to undertake it, a detailed comparison with Harvey's embryology cannot be carried out here.

     50. "... nollem dicere motum esse brutorum animam, sed potius cum Sacra Scriptura ... *sanguinem* esse illorum animam: sanguis enim est corpus fluidum, citissime motum, cuius pars subtilior dicitu spiritus, et quae ab arteriis per cerebrum in nervos indesinenter fluens totae Corporis machinam movet." (Letter to Buitendijck, 1643?, AT IV 65, CSMK III 230). Similarly, in the letter to Plemp, Oct. 3, 1637, PA I 414, CSMK III 62: "... cum Sancta Scriptura firmiter credo ... *animas brutorum nihil aliud esse quam sanguinem*, nempe illum qui, illorum corde calefactus et alternatus in spiritum, ab arteriis per cerebrum in nervos et musculos omnes se diffundit ... nescio quas animas substantiales, a sanguine, calore et spiritibus diversas, brutis affingunt." "... like the Bible, I believe, ... that the souls of animals are nothing but their blood, the blood which is turned into spirits by the warmth of the heart and travels through the arteries to the brain and from it to the nerves and muscles ... I do not see what substantial souls different from blood, heat and spirits they attribute to brutes" (CSMK text modified). Descartes refers to Leviticus, XVII, 14 ("the soul of all flesh is in its blood") and Deuteronomy XII, 23 ("blood is their soul"). Both passages occur in the context of the prohibition of the consumption of blood. On this, compare Harvey, note C 151.

     51. This was to succeed to a high degree. The opposite movement, which was soon to begin with Pascal's "logic of the heart" (Pensées, Fragments 277 and 281-283; in *Oeuvres*, 1906, vol. 13, pp. 201, 203-6), was unable to resist the dominant dualism, which was also supported by physiology. It was pushed onto the level of *projection* already sketched by

Descartes, that is, the level of the "as if", where it finally produced only a sentimental, pietistic, or erotic symbolism of the heart whose preponderance both confirmed and concealed the actual victory of Cartesian thought. On this see also, K. Weinberg, "Zum Wandel des Sinnbezirks von 'Herz' and 'Instinkt' unter dem Einfluß Descartes'," *Arch.f.d. Stud. d. neueren Sprachen und Literaturen* 203 (1966), pp. 1-31; also Morus = R. Lewinsohn, *Eine Weltgeschichte des Herzens*, Hamburg, 1959.

52. Rothschuh gives a good survey of *Descartes's* "bioergography", which is important for judging his influence on other authors: "René Descartes und die Theorie der Lebenserscheinungen," *Sudh. Arch.*, 50 (1966), pp. 25-42. The literature on Cartesianism and its influence is extensive; I have taken account of it here insofar as it focuses on the sphere of physiology. On the reception of Descartes in the Netherlands, see E. Dijksterhuis, *Descartes et le cartésianisme hollandais. Etudes et Documents*, Paris: 1951; G. A. Lindeboom, "The Impact on Seventeenth Century Medical Thought in the Netherlands," *Janus* 58 (1973), pp. 201-206; T. Verbeek, *Descartes and the Dutch: Early Reactions to Cartesian Philosophy, 1637-1650*, Carbondale, IL: Southern Illinois University Press, 1992. For England: S.P. Lamprecht, "The Role of Descartes in Seventeenth Century England," *Stud. Hist. Id.* 3 (1935), pp. 181-242; T.M. Brown, "The College of Physicians and the Acceptance of Iatromechanism in England, 1665-1695," *Bull. Hist. Med.* 44 (1970), pp. 12-30. On Scandinavia: S. Lindroth, "Harvey, Descartes and Young Olaus Rudbeck," *J. Hist. Med.* 12 (1957), pp. 209-219; idem, *Descartes in Uppsala*, Stockholm, 1964; A. Faller, "Niels Stensen und der Cartesianismus" in G. Scherz, ed., *Nicolaus Steno and his Indice = Acta Historica Sc. Nat. et Med.*, 15, Copenhagen, 1958, pp. 140-166. On the influence of Cartesianism in European science in general, E.J. Dijksterhuis, *The Mechanization of the World Picture* (Oxford: Oxford University Press, 1961) is authoritative, as is M. Boas, "The Establishment of the Mechanical Philosophy," *Osiris* 10 (1952), pp. 412-541, esp. pp. 442ff.

53. The relief and confidence that *Descartes's* new foundation for science produced in a time characterized by searching and uncertainty and by skeptical, schismatic, and esoteric doctrines is expressed in numerous contemporary statements. P.R. Sloan has followed this in his original study (Sloan, 1977). Among the authors he cites, we may take as an example Cornelis van Hoghelande (1590-1662) with his criticism of contemporary "sophists" and skeptics: " . . . their only aim seems to be, to bring into doubt . . even what is most certain and evident, to destroy all science and to spread everlasting doubt, skepticism, and, consequently, sheer ignorance through the world. But truly the incomparable Descartes has favored our century with the splendor of his mind, so that any one who follows in his footsteps easily distinguishes the doubtful from the certain, the true from the false . . . And beyond that it is the good fortune of this century, that further, what is evident and what is obscure in material things is now in general at all events apparent to every one: namely, quantity and figure . . . and local motion" (*scopus eorum unicus esse videatur, vel certissima atque evidentissima quaeque . . . in dubium vocare, scientias omnes destruere, perpetuam dubitationem et scepticismum, et conse-*

quenter meram ignorantiam, in orbem inducere. Sed profecto, saeculum quod vivimus . . . beavit ingenii sui splendore vir incomparabilis Renatus Des-cartes . . . ut qui eius vestigia sequuntur, dubia a certis, vera a falsis haud difficulter distinguant . . . Et ea porro huius saeculi est felicitas, ut insuper, quid evidens, quid obscurum sit, in rebus materialibus, in genere saltem iam cuivis pateat: Quantitas videlicet et figura . . . nec non motus localis); Cornelis van Hoghelande, Leyden, 1676, p. 135f.; first edition, 1646. Harvey, too, had to take issue with skepticism about physiological science; Sloan refers to the relevant passages (DMC, Dedication and ER 153; W. 130f.).

54. Descartes's role in the reception of Harvey is emphasized by E. Weil, "The Echo of Harvey's *De Motu Cordis* (1628) 1628 to 1657", *J. Hist. Med.* 12 (1957), pp. 167-174; the article also includes a list of negative and positive positions on the circulation. See also, Whitteridge, pp. 147-174 and Keele, pp. 153-169: "Descartes with his great reputation probably did more to promote Harvey's doctrine in Europe than any other authority" (p. 160). See also, R.S. Westfall, *The Construction of Modern Science*, New York: 1971, p. 93: "Descartes determined the tone of biological studies in the later seventeenth century far more than Harvey did . . . " The reception of Harvey in the Netherlands, also colored by Cartesianism, is described by G.A. Lindeboom, "The Reception in Holland of Harvey's Theory of the Circulation of the Blood," *Janus* 46 (1957), pp. 183-200; similarly for Sweden, Lindroth, 1957.

That Harvey's discovery was initially not at all clear in its "paradigmatic" significance is also evident in the contemporary attempts to interpret it in yet another sense: namely, in that of the Paracelsian and hermetic movements, which played a significant role especially in England. Thus, Harvey's friend Robert Fludd, physician, Paracelsist, and Rosicrucian, was the very first who took up the doctrine of the circulation and placed it in a mystical-alchemical context (*Medicina Catholica,* 1629; *Integrum morborum mysterium,* 1631). The Paris circle around Mersenne and Gassendi thus saw the two doctrines as belonging together. In addition, Harvey had had the *De Motu Cordis* printed by Fludd's publishing house in Frankfurt. In the years that followed, it was many times necessary to separate Harvey and his discovery from such an "unscientific" context, for example, in 1654 in the debate between the Paracelsian John Webster (1610-1689) and the atomist and Cartesian Seth Ward (1617-1689), who stressed the mechanistic interpretation of Harvey in opposition to Webster. On this, see A.G. Debus, "Harvey and Fludd: The Irrational Factor in the Rational Science of the Seventeenth Century," J. Hist. Biol. 3 (1970), pp. 81-105. Finally, that this "Cartesian" reading of Harvey has persisted until today can be illustrated once more in an article by J.A. Passmore, "William Harvey and the Philosophy of Science," Austr. J. Phil. 36 (1958), pp. 85-108: "Descartes and Harvey agreed that the body is a mechanical system (!)" (p. 88). "Harvey broke through a barrier; he proclaimed a unity where there had seemed to be an absolute gulf: between the animal organism and inanimate mechanical systems (!)" (p. 86).

55. Compare, e.g., the correspondence with Beverwijck 6,10 and

7,5, 1643, AT 3, pp. 682ff. and 4, pp. 3ff; see note D 26, or Professors Plemp and Vorstius (see note D 14), who questioned Descartes about certain physiological themes.

56. See, for example, R. Lower, *Tractatus de Corde*, London, 1669, pp. 60ff.; J. Mayow "De Respiratione," 1668, in *Tractatus quinque*, Oxford, 1684, p. 302; G. Borelli, *De Motu Animalium*, Leyden, 1685, p. 325; J. Bohn, *Circulus anatomico-physiologicus*, Leipzig, 1628, p. 113.

57. "Personne que luy n'a expliqué méchaniquement toutes les actions de l'homme, et principalement celles du cerveau; les autres nous décrivent l'homme mesme; Monsieur des Cartes ne nous parle que d'une machine, qui portant nous fait voir l'insuffisance de ce que les autres nous enseignent, et nous apprend une methode de chercher les usages des autres parties de la machine de son homme, ce que Personne n'a fait avant lui. Il ne faut donc pas condamner Monsieur Des Cartes, si son systeme . . . ne se trouve pas entièrement conforme a l'expérience; l'excellence de son esprit qui paroist principalement dans son Traitée de l'Homme, couvre les erreurs de ses hypotheses" (Nicolaus Steno, "Discours sur l'Anatomie du cerveau," 1666, p. 8, *Opera Philosophica*, Copenhagen, 1910, vol. II, pp. 1-35).

58. That materialistic inferences were not yet in the order of the day is clear from the opposition that Descartes's entirely logical identification of animals with automata evoked. His soulless *physiology* was acceptable, but a soulless *living being* was not yet imaginable. Cf. L. Rosenfield, *From Beast-Machine to Man-Machine. The Theme of Animal Soul in French Letters from Descartes to La Mettrie*, New York: Oxford University Press, 1940, p. 63: " . . . whereas the beast-machine aroused a furor of debate, the principle of man's bodily machine was more widely accepted." Thus, Descartes saw himself forced to call the totality of the particles of blood and spirit an "animal soul." (See note D-49.)

59. The theory of preformation, initiated by Jan Svammerdam (1637-1680) and the Cartesian Nicolas Malebranche (1638-1715), meant basically the solution of the problem of conception and development through its abolition: " . . . individual development and differentiation were reduced to growth, which could be regarded as a more or less purely mechanical process" (Gasking, p. 43). Here, too, Descartes's influence made itself felt *indirectly*: since his own embryology was unsuccessful (see note D 48), his successors looked for another way of remaining faithful to Cartesian principles. In any event, Harvey's assumption that the primordium was already a living unity even without organs was no longer acceptable for the biologists of the second half of the century: " . . . they were convinced by his (Descartes's) idea that living things must be machines and, in so far as this view played a part in the acceptance of the preformation theory, Descartes is indirectly responsible for more of the subsequent development than appears on the surface. He had, moreover, introduced into European thought the idea that a physico-chemical explanation of generation might be possible . . . The present successes in this field owe something to Descartes's heroic attempts" (Gasking, p. 69).

60. Unfortunately, there is no comprehensive study of the physiol-

ogy of the heart and circulation after Harvey. For a survey, see K.E. Rothschuh, " Geschichtliches zur Lehre von der Automatik, Unterhaltung and Regelung der Herztätigkeit," *Gesnerus 27* (1970), pp. 1-19; ibid., "Die Entwicklung der Kreislauflehre im Anschluss an William Harvey" in Rothschuh, *Physiologie im Werden*, Stuttgart: 1969, pp. 66-86.

61. Descartes lived in Holland from 1629 to 1649.

62. See K.E. Rothschuh, "Henricus Regius und Descartes," *Arch. Int. Hist. Sci. 21* (1968), pp. 39-66, with its careful investigation of the relations between Descartes and Regius, and a judgment of Regius's originality. Descartes conducted an extensive correspondence with Regius on philosophical and physiological questions (see AT II and III).

63. Cf. Sloan, p. 28. In 1640, Regius had his student Hayman defend Harvey's theory in a disputation against Primrose (*Disputatio medicophysiologica pro sanguine circulatione*, Utrecht, 1640); he himself published *Spongia, qua eluuntur sordes animadversionem qua J. Primirosius*..., Utrecht, 1640 and, in 1641, further theses in defence of Harvey (see AT III, p. 446). Regius's theory of the heart is presented in J. de Back's "Discourse on the Heart" as an appendix to *The Anatomical Exercises of Dr. William Harvey*, London, 1653, pp. 114ff.

64. Regius radicalized mechanism, and questions for the first time the independence of the human mind (*Explicatio mentis humana*... 1647); Descartes reacted with a sharp condemnation in the first French edition of the *Principles* (1647). Cf. Lindeboom (1979), pp. 25ff; also Verbeek, 1992.

65. H. *Regii Fundamenta Physices*, Amsterdam, 1646, pp. 17, 24f. (page numbers in the following given according to this edition).

66. "Actiones animales automaticae sunt, quae anima seu mente non attendente, per solum quidem organorum animalium, nempe spirituum, nervorum, cerebri, et musculorum, ab objecto interno ver externo agitatum, motum, ab homine tanquam alique automato . . . peraguntur; sed a mente tamen, si attendat, perfici possunt" (*Henrici Regii Fundamenta Medica*, Utrecht, 1647, p. 6; the page references in the text continue to follow the *Fundamenta Physices*).

67. "Influunt autem spiritus alternatim eodem modo in cordis integri vel exsecti et dissecti fibras, ut in caudis vivarum lacertarum amputatis, et in animalibus in somno respirantibus, spiritu modo in has, modo in illas fibras et musculos influere observantur"; *Fundamenta Physices*, p. 181.

68. In his *Anatomia reformata* (Leyden, 1651), Thomas Bartholinus unambiguously names Regius as the originator of this construction: " . . . *loco facultatis, spirituum animalium in fibras cordis influxum substituit Regius*" (p. 255).

69. See e.g. V. Plemp, *Fundamenta Medicinae*, Louvain, 1644, pp. 151-160. That Regius very early had a sense of the critical points in the Cartesian point of view is clear from his theory of conception, which, for all its similarity with that of Descartes (fermentation and vortical motion of the seeds fused with one another), nevertheless also displays a preformational element: the seeds of the parents contain "rudiments" of the future organism, which correspond "in rough outline" (*rudis delineatio*) to the form of the par-

ents (p. 298f.).

70. "... nulla ratio suadet, ut cordi ullam vim attracticam magneticam, aliturae causa institutam, vel aliam non intelligibilem, et superfluam attribuamus" (p. 186).

71. "... formis, facultatibus, antipathiis aut sympathiis, naturae providentia, archeis, attractione, ocultis qualitatibus, aliisve eiusdem valoris et obscuritatis principiis"; "... omnia corpora quocunque modo agentia, tanquam machinas consideranda, eorundemque actiones atque effectus ... non nisi ... secundum leges mechanicas, explicandos ... existimamus." (Cornelis van Hoghelande, *Cogitationes, quibus dei existentia; item animae spiritalitas, et possibilis cum corpore unio, demonstratur; Nec non, brevis Historia Oeconomiae corporis*, Leyden, 1676, pp. 137, 124 [page references in what follows according to this edition; first edition, 1646]).

72. "... omnes eorum (animalium) motus ... oriri ... a motus quodam particularum minimarum, extra corpus animale motarum, per visum, gustum, tactum, auditum etc., ipsorum animantium corpora (mediante ... spiritu ... animali) certo modo agitantium atque moventium, et hac ratione motus quosdam automaticos ... eis imprimentium"; p. 98f. Cf. the statements by Ivan Pavlov, note D-40.

73. "... corporis passionibus nonumquam praecedere, variosque ei motus et effectus imprimere ... Quatenus vero animalia, eosdem (homines videlicet) motibus etiam automaticis subjectos esse fatemur", p. 99.

74. "Si quis autem porro quaerat, qua ratione in partibus cordis abscissus, vel resectis, motus iste convulsivus tanto tempore continetur, cum nullus in eas influat sanguis ... praecedentium demonstrationum memor, cordis parenchyma speciali sive eminentiori fermentandi actione praeditum, sanguinisque in dictum parenchyma mediante arteria coronaria notaverit ingressum"; p. 60.

75. p. 62. In his letter to Plemp (3-23-1638), Descartes already specifies a similar experiment (the effect of the infusion of blood into an eel's heart, but without the previous extraction of blood); AT II, p. 66.

76. See Lindeboom (1978), p. 72, Rothschuh (1978), p. 279. 76 to M. Foster, *Lectures on the History of Physiology during the Sixteenth, Seventeenth and Eighteenth centuries* (Cambridge: Cambridge Univerity Press, 1924, p. 147), Sylvius deserves special credit for the spread of Harvey's doctrine in Holland.

77. "... he boldly asserted that the chemistry of living things was the same as the chemistry of so-called dead things, that what took place in a live body was the same as that which might be made to take place in a flask in the laboratory. ... To Sylvius at least is due the credit of showing that there was no ... necessary connection between chemistry and spiritualism; that on the contrary the newer chemistry in its attempts to solve vital problems trod the path of the most naked materialism" (Foster, op. cit., p. 151). On the relation between Descartes's and Sylvius's neurophysiological ideas, see Mesnard, p. 208.

78. "Disputationes Medicae" (1659-1663), in Francisci de la Boe,

Sylvii *Opera Medica*, Paris, 1679 (referred to below by chapter and section).

79. "... censemus Sanguinem in Cordis ventriculis perficiendum ab innato Ipsi Igne accendi, simulque rarefieri, atque sui rarefactione majus, majusque, quo contineatur, spatium requirere, ac proinde Cordis parietes sensim magis, magisque expandere, ac sponte tandem, ubi amplius dilatari nesciunt Cordis Ventriculi, in arterias ... irrupturum, ni Cordis parenchyma sui expansione molestatum Spiritus animales in sui vocaret auxilium, qui copia convenienti accedentes contrahunt musculos parenchyma Cordis constituentes, sicque ... sanguinem ... expellunt in Arterias" (III, 15).

80. "*Naturalem* proinde putamus vocandum Cordis Ventriculorum Dilatationem, a Sanguine rarescente factam, uti *Animalem* eorundem Contractionem per musculos absolutam, et Voluntati quodammodo parentem" (III, 16).

81. "... ab Innato quidem Cordis Igne mobilior reddatur Sanguis accensus et rarefactus; moveatur tamen et propellatur imprimis a partibus ipsum contentibus" (III, 23).

82. "... a Spirituum animalium ad Cordis Musculos motorum et potenter vel languide ipsius parenchyma contrahentium copia vel paucitate omnino deducendum" (III, 38).

83. "Primo tractabimus de homine automato, sive corpore animali, videbimusque, quaenam functiones ex artifiosissima eius structura dependeant, nihil ad eas conferente mente nostra. Secundo inquiremus in naturam mentis nostrae ... Tertio hanc mentem cum corpore uniemus ..." (Theodori Craanen, *Oeconomia Animalis, item Generatio Hominis ex Legibus Mechanicis*, Amsterdam, 1703, Preface (pages in the following acccording to the posthumously published edition). Unfortunately, I did not have access to the *Tractatus physico-medicus de Homine*; however, as I gather from Sloan's summary (p. 20), the account of cardiac activity, which is here chiefly in question, corresponds entirely to that contained in the *Oeconomica Animalis*.

84. "Ergo sanguis duo facit contraria, dilatat et constringuit; dilatat enim, quatenus est infra ventriculos; et constringit, ventriculosque angustat; sed per partem spirituum suorum, qui ingrediuntur parenchyma cordis" (p. 47).

85. *Cornelii Bontekoe Metaphysica; De Motu; Oeconomia Animalis*, Leyden, 1688 (pages from the *Oeconomica* given as in this edition).

86. "Systole primarium motus cordis est et a sangine nequaquam sed tota a musculo cordis et succo nerveo dependet" (p. 15).

87. See K. E. Rothschuh, "Vom Spiritus animalis zum Nervenaktionsstrom," p. 118f., in Rothschuh, 1969, pp. 111-138.

88. Toward the end of the seventeenth century such conceptions were supported, for example, by the experiments of Frederick Ruyschs (1638-1731) with injections of colored wax, which made visible a number of new, smaller vessels.

89. "... musculorum ventres carneos non fibris, sed tubulis, ut caetera omnia, constare, tum nerveis a nervorum filamentis, tum arteriosis ab arteria; in hos plus succi nervei et sanguine si impellitur a corde et cerebro, quid aliud fieri debet, quam tubulorum repletio, tumor et tensio?" (p.

95). The distension theory of muscular movement had really been refuted well before this by Nicolaus Steno and Jonathan Goddard. See N. Steno, *Elementorum myologiae specimen*, 1667; Goddard's experiment, which demonstrated a reduction in size of the contracted muscle by measuring its dislodgement of water during contraction, was published by Francis Glisson in his *De ventriculo et intestinis*, 1667. Nevertheless, the older theory persisted until far into the eighteenth century as it does also, despite some slight doubt (*an et quomodo intumescant . . . non facile dici potest*, p. 94), in Bontekoe. This again illustrates the power of the Cartesian thought-style.

90. "Igitur, quando musculi moventur, cor debet certo modo moveri, ut plus sanguinis ingerat in eos musculos, utque cerberum commoveat et incitet, ut eodem momento in eos musculos aliquid succi nervei indat" (p. 94).

91. " . . . ita tandem cor omnia movet, sed ipsum cor movetur a succo nerveo, in eius carnem distributo, ut hic mirabilis quidam circuitus fuit . . . (cor) sanguinem vi ac impetu in cerebrum projicit, ita ut glandulae premantur, et succus nerveus secernatur, propellaturque . . . in nervos, interque hos in nervum cordis . . . , ut simul atque contracti cordis ventriculi dilatantur, . . . succus ille nerveus in cordis carnem defluat ad novam contractionem. Atque hac ratione cor omnia movet, et se ipsum . . ." (p. 93f.).

92. p. 113. In fact, the rhythm of the pulse and of breathing strive for whole-number ratios (4:1, less often 3:1). But it is precisely the *freedom* in this congruence that is characteristic for the organism as well as oscillation around a state of equilibrium (Harvey's "equilibrium"). Cf, G. Hildebrand, "Arterielle Pulsation und rhythmische Koordination," in E. Pestel, G. Liebau (eds.), *Phänomen der pulsierenden Strömung im Blutkreislauf aus technologischer, physiologischer und klinischer Sicht*, Mannheim: 1970, pp. 34-52.

93. "Quod autem Harvaeus . . . cor existimaverit primum movere seu punctum illud saliens . . . duplex error est primo quod punctum illud . . . non sine aliis movetur, quia omnes partes simul ac semel delineantur . . . : secundo . . . omnem agitationem debent succis a matre venientibus" (p. 113).

94. "Hominis vita constitit in mentis corporisque (ut aiunt) conjunctione: sed mens vivit, et corpus quoque suam propriam vitam vivit; haec a mente non dependet, neque alia est quam circuitus motusque sanguinis ac succorum" (p. 1).

95. I am examining here Blancaard's "Institutiones Medicae" in *Steph. Blancardi Opera medica, theoretica, practica et chirurgica*, Leyden, 1701 (Page references below follow this edition).

96. " . . . cordis systole et diastole, motus intestinorum . . . peristalticus . . . . a partium structura mechanica solummodo dependent, quae ab affluentibus succis in modi hydraulici moventur" (p. 242).

97. " . . . quemadmodum alia membra, suas antagonistas habent: ita et cor, intestina etc. duplices possident fibras motrices; Nam cum hae contrahunt, illae iterum aperiunt; hoc est, quando hae sanguine implentur, illae iterum deplentur, qui motus est alternus, et *motus Peristalticus et systole et diastole vocatur*" (p. 244).

98. C. Galenus, *De Usu Partium*, VI, 8, *Opera* (Kühn), III, p. 439.

99. "But important as Harvey was in inspiring this reshaping of physiology, it is in no real sense a direct and unaltered continuation of his ideas. The new physiology was based on an atomic and chemical view of matter. Harvey, an Aristotelian to the core, was deeply suspicious of such innovations in natural philosophy and—had he lived to see the outcome—would have been distinctly uncomfortable with the results." (R. Frank, *Harvey and the Oxford Physiologists*, Berkeley and Los Angeles: University of California Press, 1980, p. xiv.) In his study, Frank describes the Oxford physiologists as a "scientific community" and investigates their theories chiefly under the aspect of the new *respiratory* physiology, in which he often points out the influence of mechanism. On the Cartesian influence in England, see also the literature cited in note D-52, for example S. Lamprecht: "Descartes . . . deeply influenced every English thinker of consequence . . . between 1640 and 1700" (p. 182).

100. "In his . . . quae sequuntur, nullum ego Harveo patrocinium parabo, cum ab eo circa illa dissentiam" (Georgio Ent, *Apologia pro circulatione sanguinis, qua respondetur Aemilio Parisano, medico Veneto*, London, 1641, p. 119; pages references below to this edition). On Ent, see Frank, p. 22f.

101. " . . . cordis parenchymati hic motus attribui non debet, sed potius sanguini in sinistro ventriculi ebullienti, a cuius intumescentia, ventriculus ab interstitio recedit, et fit diastole" (p. 117).

102. Kenelm Digby, *Two treatises, in the one of which, the nature of bodies, in the other, the nature of man's soul, is looked into in way of discovery of the immortality of reasonable soules*, Paris: Gilles Blaizot, 1664, pp. 232-238 (Reprinted by Garland: New York, 1978); quotation p. 234. Even the book's title shows its Cartesian bent.

103. Sir William Petty, "Six Physico-Medical Lectures" in *Sir William Petty's Medical Studies and Other Papers*, vol. 3. Cf. Frank, p. 92 and p. 102: "The dominant characteristic of Petty's lectures was the aggressiveness with which he preached the Harveian doctrine of the heart's pumplike motion and the resultant circulation of the bood . . . Petty's views on these questions seem to have been distinctly Cartesian, especially in his analysis of the structure of matter and his emphasis on the necessity of mechanical, deterministic explanations . . . Petty's solution, one which was to be followed by many of his friends and colleagues . . . was to expalin Harveian problems and results according to mechanistic methodologies."

104. Nathaniel Highmore, *Corporis Humani Disquisitio Anatomica*, The Hague, 1651 (page references below according to this edition); ibid., *The History of Generation*, London, 1651. It is worth noticing how Aristotelian Highmore's language can be—as with "violent motion" or "material cause"—while his explanations are plainly Cartesian. It was difficult for medical writers to change their vocabulary even while they were subscribing to a new style of thought.

105. p. 157ff. "Harvey . . . probably regarded such a concept of the blood as erroneous and would have viewed with dismay his disciple's attempt to render an analytical definition of the vital fluid in terms of heterogeneous particles" (Frank, p. 98). In his embryology, Highmore also

rejects Harvey's idea of a primary self-motion of the blood: what appeared as a pulsating dot of blood was seen under the microscope to be an already beating heart (Frank, p. 101).

106. "Tertius vero spirituum motus hic in corde observatur, qui licet a spirituum animalium affluxu dependeat, non tamen in nostra potestate est . . . naturalis vocatus, non quod a spiritibus aliquibus praeter animales fiat; sed quia natura sine illo subsistere nequeat" (p. 137).

107. Foster remarks on Willis's views: "They are indeed to a large extent the views of Descartes, modified by more exact anatomical knowledge, occasionally by sound physiological deductions. . . . He admits with Descartes that man possesses a rational soul, an immortal, incorporeal soul, but that, putting aside everything which is due to the direct activity of this rational soul, the nervous as well as the other phenomena . . . may be regarded as the phenomena of a corporeal machine" (op. cit., p. 268). Cf. also Frank, pp. 165ff., 232ff.

108. Thomas Willis, *Diatribe duae Medico-Philosophicae: De Fermentione; De Febribus* (1659), 4th ed., London, 1677, p. 3 (page references in what follows to this edition).

109. "Fermentatio est motus intestinus particularum, seu principiorum cuivis corporis, cum tendentia ad perfectionem eiusdem corporis, vel propter mutationem in aliud" (p. 17).

110. "Sanguini . . . insunt *particulae heterogeneae*, quae cum diversae sint figurae, et energiae, earum mutuo occursu, et subactione, quamdiu in mixtione perstant, fermentationis motus jugiter conservatur" (p. 99).

111. In his *Apologia,* Ent had mentioned for the first time the "nitrous air" or "nitrous virtue" of the respired air (*Apologia*, p. 97f.) The latter expression also points to the source of this idea in Paracelsus, who had seen the same effective principle in lightning (and thus in the air) and in gunpowder. Saltpeter had also long been known as a preservative for organic substances and as a source of energy for reactions in the laboratory (see Frank, pp. 118ff.). In the seventeenth century, the idea spread more and more widely that it could be a question here of the material connecting link between fire, air, and life (e.g. also in Sylvius, *Disputationes Medicae*, VII, 77f.). Thus, we have here, to speak with Ludwik Fleck, the "pre-idea" of oxygen; to reach that point, however, there was still a long series of metamorphoses to be traversed. In Oxford, this pre-idea was first stripped of its "virtue" character (by Ralph Bathurst, among others, 1654), and assimilated to the corpuscular philosophy, in order finally to reach the provisional high point of its development through John Mayow.

112. Thomas Willis, *Cerebri Anatome*, London, 1664 (page references according to this edition).

113. "In spirituum animalium diataxi et ordinatione . . . Animae sensitivae . . . essentia constitit, qua nempe solummodo est spirituum illorum (qui velut Atomi, seu particulae subtiles sibi mutuo adhaerentes et concatenatae, sub certa specie configurantur) systasis quaedam . . . " (p. 132).

114. p. 104ff. This principle, that the former functions of the soul are to be localized in various centers of the brain, has been a guiding prin-

ciple for the further development of neurology. Concerning Willis on this point, see Foster, pp. 270, 272f.

115. "... in Corde spiritus animales, intra fibras et filamenta nervosa ... scatentes, a sanguine influo ... corpuscula sulphurea copiose suscipiunt; quae dum iidem spiritus ... excutiunt, ac veluti explodunt, totius Cordis (lateribus eius cum impetu quodam intra deductis) systole infertur, qua quidem sanguis ex utroque sinu velut emboli impulsu ejicitur" (p. 141).

116. This designation occurs in Willis's later work, *De sanguinis accensione*, London, 1670, p. 64f. (cf. Frank, p. 236).

117. Richard Lower, *Tractatus de Corde*, London, 1669, pp. 60-72 (pages references according to this edition).

118. "Tantum etiam abest ut credam sanguinis motum a sua in Corde accensione ulla dependere, ut nec Cordi calorem suum jure aliquo debere videatur" (p. 71).

119. "... vesicula ipso sanguine prior erat ... ab ipso principio, liquoris in sanguinem mutati non tantum conceptaculum, sed et machina motiva est" (p. 69).

120. "Quod ad undulationem istam sanguini in vena cava post emortuam auriculam; arbitror illam nullo sanguinis intestino motu, sed vasorum, a spiritibus per nervos ubique distractis, corrugatione contingere: Non aliter quam spiritus in musculis ... motum illum tremulum post mortem diu protrahunt" (p. 70).

121. "Cum ... constiterit motum Cordis non dependere a sanguine, proxime dicendum restat, quibus instrumentis et machinis perficiatur" (p. 74).

122. "Uti vero sanguinis et cordis motus a cerebro totum dependet ... cerebrum ipsum in partes omnes inferioris corporis velut rex in suos subditos dominatur, et pro nuto et imperio suo omnia regit et gubernat" (p. 92f.)

123. Nicolaus Steno, "De musculis et glandulis observationum specimen," 1664, p. 181f., *Opera Philosophica*, I, pp. 161-192 (from the German version of K.E. Rothschuh, 1968, p. 58). Steno's Cartesian training is still evident in the spirit of these sentences. Although his distinguished anatomy of the musculature also influenced the physiology of the heart, I have not dealt separately with Steno's works here since he did not really explain the movement of the heart or set it into a physiological context.

124. P. 169f. On this complex see also, Frank, pp. 195ff., 213ff.

125. John Mayow, *Tractatus Quinque Medico-Physici*, Oxford, 1674, including "De Sal-nitro et spiritu nitro-aereo" (I, pp. 1-265; cited as S); "De Respiratione" (I, pp. 267-308, cited as R); "De Motu Musculari" (II, pp. 1-106, cited as M). On Mayow, see Frank, pp. 258-274; Foster, pp. 183-197, as well as W. Böhm, who calls Mayow the "enthusiastic disciple of Descartes" (p. 46). Rothschuh sees him somewhat more cautiously as "strongly influenced by Cartesian natural philosophy" (Rothschuh, 1968, p. 117). Here, we can only sketch in a simplified survey Mayow's complex system of chemistry and phys-

iology, which was founded on countless experiments and careful inferences.

126. "Aerem hunc circumfusum, qui tenuitate sua oculorum aciem fugiens, ipsum perpendentibus spatium quasi inane apparet, sale quodam universalis, indolis Nitro-Salinae, Spiritus sc. vitali, Igneo summaeque Fermentativo impraegnari ex sequentibus ... manifestum erit" (S p. 1).

127. "Vita, ni fallor, in spirituum animalium distributione consistit; quibus supplendis cordis pulsatione, sanguinisque ad cerebrum affluxu omnino opus est: et videtur respirationem ad cordis motum ... praecipue conducere. Enimvero verisimile est, ad quemvis musculorum motum sal hoc aereum omnino necessarium esse; ita ut sine eodem neque cordis pulsatio fieri possit" (R. p. 304).

128. Böhm has investigated in more detail this development of Cartesian ideas and principles in Mayow's system. According to him, in the case of the saltpeter ("oxygen") theory, it is a question above all of a speculative extension of Cartesian cosmology and physiology (Böhm, p. 65).

129. "Contra praedicta in promptu est objicere, spiritus Animales praecipuam corporis partem constituere; eoque probabile esse eosdem non ab aere, qui quid extraneum, et a corpore alienum est, provenire, sed potius a particulis sanguinis nobilioribus ... Porro machinae animalis artificio plane admirando minus convenire, ut eadem a principio externo in motum concitetur" (M 44f.).

130. "... semper Erasistrati institutum aestimavimus, omnia quae in corpore nostro contingunt Mechanice explicare ... Eas vero machinas esse statuendas, quas evidens ratio et ... sensus ostendant." Jan de Wale, "Epistolae duae: de motu chylis et sanguinis ad Thomam Bartholinum," in Caspar Bartholin, *Institutiones Anatomicae*, Leyden, 1641; quoted here from Thomas Bartholinus, *Anatomia Reformata*, Leyden, 1651, pp. 529-576; above passage, p. 562.

131. Hermann Conring, *De Sanguinis Generatione et Motu Naturali* Leyden, 1643; quoted here from the edition of 1646.

132. See note 129 (pages according to edition specified there).

133. "Abhoc sanguine continuo movetur cor, instar rotae molendariae, perpetuo aquarum insultu agitatae, quo intercepto cessat. ... Unico experimento fidem assertioni facio: si vinculo intercipiantur vasa cordis adferentia, motus cordis desinit, reliquusque erit undulans ... " (p. 251). In the *De Motu Cordis,* Harvey describes the same experiment.

134. "*Facultas pulsifica* cordi insita admotum eiusdem cum sanguine necessario est conjungenda, sive ut sanguini influxum et exitum juvet dirigatque, inordinate alioquin procedentem ... Pulsus foret inaequalis semper, quia influxus inaequalis, nisi a facultate dirigeretur" (p. 255).

135. My previous argument is confirmed by the fact that Bartholinus unambiguously identifies the theories of the external regulation of the heart (through the cardiac nerves and the animal spirits) with the Cartesians (Regius and Hoghelande). The English physiologists omit the corresponding references; the opponents of a point of view are often more clearly aware of its origin than are its proponents.

136. "... sanguis copia sua movit cor; et cor, constrictione sua

propellit sanguinem; et ita utriusque motus perennatur." "Pulsum a duplici Agente peragi, sanguine nimirum, et corde ... hac mutua operae societate alternatim institutae, sanguis per totum corpus propellitur ... Pulsum cordis a sanguine, ad motum contractionis irritante, oriri." Walter Charleton, *Exercitationes Physico-Anatomicae de Oeconomia Animali ... Mechanice Explicata.* Amsterdam, 1659, p. 105, p. 108. Cf. the same statements in Harvey; see above, n. C-108.

137. Cf. Pagel (1967b), p. 499f. Quotations below from Francis Glisson, *Tractatus de Ventriculo et Intestinis*, London, 1677.

138. "Naturalis enim perceptio (seu Archeus) innumera prope novit, quae sensum latent; et totam fabricam corporis, quam ipsa formaverat, et usum partium modumque quo omnia peragenda sit, callet" (p. 181).

139. "Cordis pulsatio nec sensu peragitur, nec variatur. Fibrae cordis, virtute micationis vitalis sanguinis in eius Ventriculis contenti, per vices irritatae, excitantur ad se contrahendas et pulsationem faciunt; mox Irritatione remissa relaxuntur, et naturalem partium positionem repetunt. Negari non potest evidentem hic fieri fibrarum irritationem. Secundam hanc enim rythmus pulsationis variatur, ut ex pulsuum differentiis in febribus et aliis morbis constet (p. 170).

140. "Sed inquiunt, naturam audere et posse aliquid supra leges mechanicas. Egregie profecto, quasi leges mechanicas non essent necessariae, et proinde natura contra leges necessitatis, a divina sapientia praescriptas, scilicet impossibilia agere posset." Giovanni Borelli, *De Motu Animalium* (probably written for the most part about 1660), Rome, 1680-1681. Quoted here according to the Leyden, 1685 edition, Part II; above quotation, p. 22.

141. " ... immediata causa tensionis cordis, erit inflatio ... pororum eius, facta a fermentativa ebullitione tartarearum partium sanguinis a succo spirituoso, ex orificis nervorum instillato" (p. 110).

142. On page 325, Borelli goes in detail into the views of Descartes and the Cartesians.

143. P. 112f. A Cartesian idea could have been godfather to this view as well: according to Descartes, the blood enters the chamber of the heart in the form of a large *drop* and so triggers the movement (*Discours*, V, AT VI, p. 49; TH, AT XI, p. 123).

144. "Cumque sit incredibile, ut cor modo ab una, modo ab alia diversa causa motiva agitetur, erit consentaneum, ut etiam consuetae cordis motiones fiant non necessitate mechanica, ut in automate, sed ab eadem animastica potentia sensitiva ..., quae grandem illam pulsationem in gaudio, et minimam motionem in timore efficiebat" (p. 115).

145. "Quare non videtur omnino ineptus, et riso excipiendus is, qui dubitaret, pulsationem musculi cordis non a caeca necessitate mechanica, sed ab eadem facultate animali, habitu quodam, fieri posse sine advertentia, a qua omnes artuum musculi arbitrariis praeceptis moventur" (p. 116f.).

146. Johannes Bohn, *Circulus anatomico-physiologicus, seu Oeconomia corporis animalis*, Leipzig, 1686 (page references according to this edition).

147. "Neque etiam omnes corporis humani motus ab eadem anima cognoscente, sed interdum a solis objectis aut cerebri peculiari dispositione

... determinantur ... Adeoque diversi motus, juxta *Cartesium*, equidem in homine, non tamen per hominem, exercentur"; " ... principium motus materiale et mechanicum in nobis perinde ac in brutis, supponamus" (p. 46).

148. "Quod enim machinae hydraulicae antlia seu embolus primarius, illud in machina animali huiusque fluido catholico circumagendo, praestat cor: quatenus ille aquae, hoc sanguini, evehendo ac propellendo, primum impetum imprimit, utroque autem fatiscente utriusque fluidum stagnat" (p. 99).

149. " . . . immo supponere licet moveri ab eadem necessitate mechanica seu organica, scilicet adaptatione atque tumefactione fibrarum, deinde etiam per spiritus animales" (p. 103).

150. " ... qualibet fibra vitalis instinctu quodam gaudet, per quem ad sui contractionem seu ad motum stimulatur; quem stimuli sensum seu perceptionem, si non tactus speciem, certe huic analogam et determinationem ad motum censere convenit ... Pariter nec cordi suum stimulum, per consequens nec sentiendi vim, deesse, ex superius dictis innotescere potest ... " (p. 449f.).

E

VITALISM AND MECHANISM BETWEEN 1700 AND 1850

Summary. *After a backward glance at the first phase of the reception of the circulation, we will follow the interaction of the vitalistic and mechanistic aspects in the period from 1700 to 1850. Within a brief time in or around 1750, after a period of increasingly disputed dominance, the mechanistic paradigm was replaced by vitalism. Above all, vitalism was shaped by the concepts of "irritability" and "sensibility," through which—without being aware of it—Albrecht von Haller had again taken up Harvey's conception in modified form. An essential contribution to this development was the inability of the mechanistic orientation to provide a satisfactory explanation of cardiac action. Haller demonstrates conclusively its independence of the central nervous system and, like Harvey, interprets it as an autonomous stimulus-response reaction of blood and cardiac muscle. The vitalism that was attached—contrary to his intention—to Haller's results and concepts not only takes up again Harvey's explanation of cardiac action, but also his basic morphogenetic point of view. The attempt is made to deduce the functions of the organs from their development, and the movement of the heart in particular from the primary, autonomous movement of blood in the egg (C.F. Wolff). Once more, the blood is often considered the sensitive basic material of the organism, itself alive.*

In a further radicalization of the vitalistic approach, the physiology of the Romantic period (from about 1800) then contrasts the heart as mechanically acting center with the periphery *as the proper sphere of vital processes. In 1801, Xavier Bichat attributes the circulation itself to a "peripheral heart" to which the central heart is subordinated as a supplementary stimulus. Under the guidance of the motif of* polarity, *indicated by Harvey but now made explicit in the language of* Naturphilosophie, *German physiologists now also develop various models opposed to the previous conception of the circulation (P.F.v.Walther, L.Oken,*

J.H. Oesterreicher, C.H. Schultz, C.G. Carus).

About the middle of the nineteenth century, a new thought-style again comes to dominate, a thought-style that is related, despite modifications, to the Cartesian-mechanistic tradition. With his exposition of the movement of the heart and blood, F. Magendie can serve as typical representative of this conception. Nevertheless, in the sense of the interaction of different aspects, the new paradigm does not fail to be influenced by its predecessor. In conclusion, this is illustrated by the explanation of cardiac action established in about 1850 (and still valid today), in which "Romantic" and mechanistic theories have converged.

We have traced the influence of Cartesian thought on the physiology of the heart and the circulation for more than half a century. In scarcely any other sphere do we find so evident the fundamental upheaval of physiology through a physicalistic approach and the removal of the soul from the body, as Descartes had effected it, certainly not alone, but still in paradigmatic fashion.

Thus, the heart loses almost all the functions it had previously had. Above all, it is no longer the life-giving organ of the body. With the discovery of the circulation, Harvey had already deprived it of the task of forming the blood in cooperation with the liver. Since the spirits had been reduced to the animal spirits—a reduction suggested by Descartes's emphasis on the animal spirits—the heart serves only as stimulating organ for the spirits en route to the brain. Finally, after the reinterpretation of vital heat as an organic-chemical reaction of particles, the heart for a while plays the role of a "reaction flask"; but with the shift of the energy-giving process to the arterial blood, this function, too, was lost. Nor, in this process, does the blood inherit the heart's function as "life stuff" in the way Harvey had regarded it; rather, as a mixture of particles, it becomes the purely material bearer of atmospheric and nutritional components or of their "energy."

The heart no longer carries the vitality of the organs—and it loses its own vitality. Neither a pulsative faculty nor any other kind of autonomy is allowed it; it works, like the other muscles, as an executive organ for the brain in accordance with the central disposition of the spirits. In other words, the "part" is not related, in autonomous exercise of its own function, to the "whole" of the organism, but becomes rigidly bound to this whole as determined by an external agency. Parallel with this development, finally, a further function of the heart is relegated to a merely derived, secondary role: from being the *primary* seat of emotional life, the heart becomes merely its *apparent* center. The sensitive soul is materialized in the movements of the particles of the spirits; the sensations of feeling that are thus engendered are projected into the region of the heart by the rational soul in correspondence with the paths of the nerves.

In compensation, as it were, the single remaining function of the heart becomes an unlimited, absolute one: as a mechanical pump, it governs the circulation of the blood; the streaming of the blood is now the product solely of its power of contraction and for the rest is subject to the general laws of physics and hydraulics. From an embryological point of view as well, the heart is no longer formed from the movement of the blood, but as a preformed organ, already controls this movement from the beginning.

And yet, on the other hand, the heart, among all the organs, seems to have offered most resistance to this dualistic transformation of physiology. As we have seen, a number of physiologists, even among those who were mechanistically oriented, considered the current explanation of cardiac activity unsatisfactory. The alternative consisted, on the one hand, of granting to the intellectual soul, now thought to be uniform, the additional responsibility for the vital functions; but on the other, of developing, in continuation of Harvey's conception, the notion of irritability and sensitivity, and thus of a *prepsychic vitality* of the heart. However, neither of these ideas could be smoothly united with the dualistic vision of human nature. Hence, the movement of the heart represented one of the decisive points of dispute in the emerging conflict between mechanistic and vitalistic tendencies.

Thus in the first half of the eighteenth century, the mechanistic point of view was at first still dominant. Above all, *Friedrich Hoffmann* (1660-1742) in Halle and *Herman Boerhaave* (1668-1738) in Leyden, among the most significant and influential physicians of this period, continued the Cartesian tradition. In the *Fundamenta Medicinae* (1691), Hoffmann treats the heart as the hydraulic machine of the body[1]; it is no longer the "workshop" of the blood, but only its mechanical pump (*nam tamen officina, quam antlia sanguis, ad illum . . . expellendum inserviens*; p. 22). Instead of by an *"insita vis motrix"*, an innate motive force, it is moved by the united "force of the expansive ether, the animal spirits, and the blood." The provenance of the spirits from the cerebrum or the cerebellum distinguishes voluntary from involuntary movements (p. 40); the determinate structure of the nerve streams constitutes what is also called the "sensitive soul" (*anima sensitiva*; p. 33).

In his famous *Institutiones Medicae* (1708), Hermann Boerhaave assigns the same cause for both kinds of movement.[2] But he gives a new mechanical explanation for the rhythm of cardiac action. Since the vagus nerves take their way between the aorta, the pulmonary artery, and the auricles, the heart interrupts its supply of spirit through these nerves and is paralyzed. Hence it goes limp, the arteries are constricted, setting the nerves free again, and the game starts over again (secs. 189f., p. 409). Boerhaave has thus found one of the most original interpretations of cardiac action; nevertheless, he cannot entirely free himself of the impression of an "occult" quality of this lasting rhythmical movement (*mirifica, et occulta est in corde fabricato proclivitas in reciprocandas systoles et diastoles*; sec. 187).

Up to the middle of the eighteenth century, these and similar mechanistic theories of the heart were clearly predominant. This is demonstrated not only by such mechanists as we have been considering, but by their critics, for example *Francois Boissier de Sauvages* (1706-1767) in Montpellier. In his *Dissertatio Medica de Motuum Vitalium* (1741), he turns against what he considers the commonest cardiac theory of the day with the argument that the movement of the heart engendered by the spirits amounts to a *perpetuum mobile* (heart>brain>heart). Only the soul, which induces the expulsion of the contents of the heart—felt to be disturbing—could also renew again and again the force necessary for this process.[3]

In this attitude, Sauvages and others agree with *Georg Ernst Stahl*, the most outspoken advocate of a counter-movement to Cartesianism that is gradually developing. According to Stahl, the simple, wise, and unconscious soul is itself the immediate cause of bodily processes. It needs neither faculties nor spirits in order to set the parts of the body, including the heart, in motion.[4] But from this, it also follows that the parts possess no kind of autonomous activity and, considered on their own, again stand in a purely mechanical relation to one another. *Thus Stahl's animism, even in its opposition, only strengthens Cartesian dualism; and it substitutes for the centrality of the brain the centrality of the soul.*[5] He was unable to break out of the dominance of the mechanistic paradigm, which had been strengthened by the triumphal march of Newtonian physics. Its solution to the problem of the heart remained, in the end, unsatisfactory; yet in physical haemodynamics, it still provided some significant results (Stephen Hales, Daniel Bernoulli).[6]

What led to a change of course was the pursuit of the second of the two alternatives mentioned above when *Albrecht von Haller* (1708-1777) and his students took up again Harvey's and Glisson's ideas of an irritability and prepsychic vitality of the tissues and provided for it a developed experimental foundation. Thus, in his *Dissertatio Physiologica de Irritabilitate* (1751), Haller's student *Johann Georg Zimmermann* (1728-1795) arrived, on the basis of his own experiments, at the following conclusions:

> I have found that irritability is effective above all in those parts whose movement appears necessary to life in the highest degree ... I am continuing in the investigation of the causes of movements that persist after death. We cannot relate them to the nerves. ... If one wants to take refuge in the claim that animal spirits which are still active in the muscles bring about this effect, we have no place of refuge there, no little place where they are stored; for if the spirits are used up, there is no place available here that prepares new ones. ... This reflection alone *disproves* ... *the notion of the control of the spirits over the isolated parts* ... Should the cause of these movements be sought in the soul itself, or the cause of all vital and natural movements whatsoever? ... Stahl asserts that the soul, with the help of the tonus, ... moves the blood and all the fluids; indeed, by body itself his adherents understand nothing but an *inert state* ... [But] an excised heart beats. Is there not therefore in the body another principle of movement besides the soul? Is it not evident, for example, ... from these experiments, that heart and intestines *conceal in themselves the principle of their movement?* ... Thus irritability deserves to be counted among the primary characteristics of bodies, which is common to all animal organisms and *alone, so to speak, brings about their life*.[7]

Haller himself, who preferred to distance himself from vitalistic tendencies, did not go as far as his student. Still, he too takes as the basis of cardiac movement an autonomous irritability—and that means for him *contractibility*, although he does not, like Glisson, combine this with a "natural perception." The cardiac fibers react to the stimulus of the blood by contracting, and that removes the cause of the irritation.[8] With the interruption of the entrance of blood, the heart comes to rest; the influx of air or fluid sets it once more in motion. The heart is endowed in especially large quantity with the muscular "innate force" or "irritability," for which reason it retains this power for some time after death. In contrast, the cardiac nerves can only modify the movement of the heart—during life; Haller again clearly separates the involuntary from the voluntary, neurological induced movements.[9]

Haller's conception of irritability is much reduced compared to Glisson's. Sensation and "appetite" are lacking (that is, inclination and aversion with respect to stimulation); irritability as the power of shortening is confined to the muscles; stimulus and reaction become a quasi-mechanical sequence. This and the isolated examination of vital parts allow "anatomia animata" to carry in itself once more the seed of a future "anatomia inanimata." All that is still lacking is the material substrate of the "innate force" (*vis insita*), about which, nevertheless, Haller already speculates.[10]

Yet contrary to his own intentions, Haller's experiments at first gave considerable incentive to vitalistic movements since they suggested specific qualities of the living, which were not explicable within the framework of the mechanistic paradigm. The aspect that had been latent since Harvey's time now won the upper hand, and in a short time, vitalism replaced mechanism as the dominant paradigm. The *forces animées*, the *principe de vie*, among the French physiologists, the "Lebenskraft" or "essential force" (*vis essentialis*) among the Germans, took the place of iatromechanical explanations and also altered the theories of the function of the heart and circulation.

On the one hand, this concerned cardiac activity itself, which was interpreted increasingly in terms of irritability and sensitivity, decoupled from the nervous system, and related to the blood. Thus, *Felice Fontana* (1730-1805) modified Haller's explanation by claiming that the heart loses its irritability for a short time with every contraction and must thus go limp diastolically even under continuing stimulation (*De legibus irritabilitatis*, 1763).[11] With this discovery of the *refractory phase*, the continued beating of the excised heart could be explained as the effect of a persistent stimulus from the environment (air, heat); for now, the *change* of the two movements of the heart was grounded in the nature of its own irritability—a step toward the later localization of an independent center of reaction in the heart itself.

Now, on the other side, in consequence of the "revalorization" of the movement of the blood as the preordained releaser of cardiac action, the absolute *control of the circulation by the heart* was called into question for the first time. The question of the primacy of heart or blood was raised again in a new form: was not the *periphery* to be seen as independent, indeed, as primary, and as having priority over against the merely mechanically (re)acting center? Above all, at the beginning and at the end of life, signs of a primary vitality and movement of the blood were observed—as they had been by Harvey, no longer, however, in the form of a central pulsation, but of a peripheral flow.

After excision of the heart in frogs, Haller had already found a partly oscillating, partly retrograde flow of blood in the arteries and an orthograde flow in the veins—movements that lasted up to thirty minutes.¹² Neither gravity nor contraction of the vessels could be the cause of these phenomena, according to Haller, since the blood moved in different directions and in the mesentery even *outside* the vessels. Haller attributed this remarkable "life without a heart" (*haec absque corde vita*, 237), on the one hand, to a kind of "surface tension" of the blood in the membranes of the vessels, and on the other, to an inner, "magnetic," attraction of its "globules" (erythrocytes), which makes the blood always collect wherever its great mass first rests, that is, in the larger vessels.¹³ Harvey had interpreted the centripetal tendency of the blood during life (*intra vitam*) in the same way; we have met similar reflections also in Hermann Conring.

The counterpart of these movements was observed by *Caspar Friedrich Wolff* (1734-1794) in embryogenesis: a primary flow that unites the islands of blood arising in the *periphery* of the egg, even before the beginning of the heartbeat (*Theoria Generationis*, 1759):

> ... so, indeed, we see that this movement of the fluids, through which the vessels are formed, cannot possibly be ascribed to the heart. On the contrary, it is clear that a force occurs here which is no different from that force in plants, and to which I have therefore given the same name of essential force (vis essentialis).¹⁴

Inappropriately, Wolff declares, it is customary to "derive not only the movement of the blood, but that of all other fluids . . . from the movement of the heart." (p. 268) This is indeed necessary for the distribution of blood in the adult organism; yet this movement only overlies the *vis essentialis*, which is effective both early and late:

> The first causes of the arrangements that are already present in the embryo never stop being effective; even in the adult it is they that contribute most to the processes involved ... I have called . .. the apparent causes that appear to be so important in the adult *causas accessorias* . . . They are agreed to be mechanical . . . .¹⁵

In the period that followed, a renaissance of the blood as "life stuff," as an organic body endowed with it own irritability and mobility, was based on the observations of Haller and Wolff, on the non-coagulability of the blood in the vessels, on its obviously heat-producing properties, and so on. This was the case, for example, with *John Hunter* (1728-1793) or *Georg Levison* (? - 1797).¹⁶ *Karl Friedrich von Kielmeyer*

(1728-1793), in his division of the organic world according to its respective dominant forces (sensibility, irritability, reproduction, the power of secretion) ascribed to the blood a "power of propulsion" —a concept which came to be commonly used to refer to the *autonomous movement* frequently ascribed to the blood and contrasted with the central impulse of the heart.[17]

All these vitalistic tendencies—which were occurring in parallel fashion in France—culminated in the work of *François Xavier Bichat* (1771-1802) in a comprehensive critique of the usual view of the process of circulation and the attempt to provide a "decentralized" interpretation. According to Bichat's conception, developed in his *Anatomie Générale* (1802), the influence of the heart reaches only as far as its pulse, that is, to the end of the arteries; its power breaks down in the branchings of the immense realm of the capillary flow. Here, we find the beginning of an entirely new region whose vital processes and movements are no longer subject to the central rhythm, but are founded on an "organic sensibility" of the tissues.[18] Not the heart, but the capillary system represents the end point and the new beginning of the movement of the blood (p. 247): the "tonic forces" of the tissues and the "insensible contractility" of the capillaries restore its motion (p. 426, 509). Hence, there are, at bottom, two opposed circulatory systems, above all also qualitatively different: the "red blood system" begins in the pulmonary capillaries and ends in the periphery of the body; the "black blood system" takes the opposite course. Bichat compares them with the roots, trunks, and crowns of two trees growing in opposite directions—in both chambers of the heart the "trunks" (that is, the large central vessels) are connected with one another only externally and mechanically.

Thus, over against the distinction usual since Harvey between the "major" and the "minor" circulation, Bichat places the "Galenic" conception of two blood systems that exist alongside one another— even though at the ends they pass over into one another—two systems between which the heart is inserted in order to accelerate the process since the "vital forces" of the capillary regions are insufficient for the whole path (p. 250f.). Bichat supports with various pieces of evidence his thesis that the circulation is not an achievement of the heart, but proceeds from the periphery—among others, with the fact that an adequate circulation is maintained despite serious diseases or defects in the heart; or that the blood continues to flow when the heart has stopped beating; or the phenomenon of the portal vein circulation or the circulatory flow of blood in lower animals, which sometimes proceed without an intercalated heart (p. 252f.):

Let us then stop considering this organ as the sole agent that presides over the movement of the large vessels and over that of the small ones, which produces .... inflammation in the latter, which by its impulsion causes the various cutaneous eruptions, the secretions, the exhalations, and so on. The whole doctrine of the mechanists rested, as we know, on this extreme extension that they had given to the heart for its movements.[19]

The new interpretation of the circulatory process—whether on the basis of a primary autonomous movement of the blood or, as with Bichat, of a "peripheral heart," fell, above all, in the *physiology of German Romanticism* on fertile soil since it corresponded to the basic vitalistic tendency of that movement. In addition, Schelling's concept of *polarity*, which was characteristic of the Romantic movement, found an essential application and confirmation in the contrariety of the center and periphery in the circulation. In his *Physiologie des Menschen* (1802), which was programmatic for the Romantic school, *Philipp Franz von Walther* (1782-1849) writes:

... those who think that the mass of blood is driven forward through the forces of the blood vessels in a mechanically communicated movement have really transformed the organism ... into the most perfect hydraulic machine. The whole vascular system with heart and vessels is only the skeleton of the circulation and receives its life from the blood that flows in it ... The veins would ... certainly not be capable of driving the blood upward against its own weight, if it (the blood) did not have in itself the striving to complete the other half of its circular movement in the backward direction. Therefore the circulation is not governed by any single law of hydraulics.[20]

Lorenz Oken (1779-1851) relates polarity to the newly discovered animal electricity:

The circulation has therefore three factors, the lung as oxygen pole, the capillaries as hydrogen pole, the blood as indifferent water. The circulation is a galvanic process ... (it is) the consequence of dynamic powers, not of mechanical arrangements. The heart beat is not the cause ...

> but rather, on the contrary, ... the consequence of the circulation.[21]

On the other hand, *Ignaz Doellinger* (1770-1841) or *Johann Christoph Reil* (1759-1813), for example, retain Bichat's conception of an autonomous irritability and contractility of the peripheral vessels:

> If we do not ascribe to the blood vessels ... a peculiar power of movement, we can hardly explain on the basis of the mere force of the heart and the larger arteries the constant and uniform movement of the fluids.[22]

Johann Heinrich Oesterreicher (1805-1843), a student of Doellinger's, once more summarizes the arguments mentioned by Haller and Bichat against the "sole rule of the heart" and, beyond these, points to the independence of the lymphatic system as well as to the circulation of the blood in teratological births lacking a heart.[23] His argumentation resembles Harvey's reflections on the primacy of the blood:

> If we consider that what is contained is not there for the sake of the container, but that, quite the contrary, the container must be there for the sake of what it contains, it is, if only for that reason, incomprehensible how, in looking for the causes of the circulation, any one could ever think only of the vessels and the heart, and take the blood for a meaningless accident that would just be pushed around (p. 188).

Oesterreicher subscribes to John Hunter's view "that the blood in itself is alive" (p. 189). But its primary movement occurs under the activating influence of the *nervous system*: the spinal cord as its central organ, which sends out the nerves as "rays," is related to the blood as the sun is to the earth—in other words, as unmoved but moving center of the circulation. Thus, Oesterreicher comes to the conclusion: "The fundamental cause of the circulation of the blood lies in the blood, because it is alive, and in its relation to the marrow of the nerves; the heart is only the "mechanical auxiliary device" for this movement.[24]

Again, compared with Bichat, we find a modified conception of the circulatory process in the case of *Carl Friedrich Schultz* (1798-1871). According to Harvey, Schultz remarks in his *System der*

Circulation (1836), the heart itself is a point in the circulation of the blood: "the blood does not run around the heart, as around its center, but . . . through the heart, as through every other point of this periphery."[25] In contrast, Schultz develops the idea of a "peripheral circulation" in the area of the capillaries in opposition to which the heart develops as a true *center* (*Mittelpunkt*):

> In relation to the heart, the peripheral system always remains relatively independent, and internally coherent in all organs of the body, though in differing degrees. Between these organs and the heart, connecting vessels then arise, vessels that connect the heart with this periphery, as the radii of a circle unite the circumference with the center; these are the trunks proceeding from the heart, which ramify in the periphery. (p. 254)

The peripheral circulation without a central organ appears in the lower phylogenetic stage as well as in the embryonic state; the heart develops from the movement of the blood, at first as a vessel, and in this process, "tangentially" to this central movement, the circulation is maintained *around* the center (p. 165, p. 191). The fact that Schultz explicitly opposes his morphogenetic interpretation of the circulation to Harvey shows how far Harvey's genuine point of view had been forgotten under the influence of its mechanistic reception:

> Harvey started . . . from the complete structure of the developed heart . . . , found in it the essential mechanical conditions for the movement of the blood and then believed he could explain all the vital operations of the circulatory system from this mechanism . . . (But) it is precisely what was presupposed in this, namely the developed anatomical structure of the heart, whose very development must first be understood through the vital process. While the movement of the blood is already in existence, the heart, with all those anatomical conditions, is formed . . . through the shaping power of the blood . . . . Thus it is impossible that the anatomical structure of the heart . . . should itself be the cause of this vital process. . . . The mechanism of the circulation . . . is really nothing but a *means*, through which the circulatory process . . . produces its true ends (p. 3f.).

As evidence for the independence of the peripheral flow, Schultz adduces, in addition to the arguments already mentioned, the beginning of this movement *before* the heart beat in many hibernating animals; as well as experiments that produced a persisting flow of blood in excised organs like the spleen even after ligature of the vessels (p. 268, p. 294f).[26] In the explanation of this autonomous activity, Schultz says, one must "free oneself entirely" from the usual derivation from "general physical or chemical causes" (p. 295):

> Thus if the context of the phenomena is considered as a whole, then the forces of the movements lie . . . originally in the vital attraction of the blood through the peripheral system . . . and in the simultaneous repulsion from the peripheral system . . . into the roots of the veins (p. 319).

So, we find here again the Galenic "attraction" and "repulsion" of the periphery that Harvey had eliminated: according to this view, attraction depends on the entry of plasma into the tissues so that new blood is attracted (thus among other things, the circulation of the portal vein is to be explained by the suction of the liver is to be explained on the basis of its secretion of bile); repulsion arises as increased "tension between blood and vessel" through the secretion of metabolic waste into the blood of the capillaries.[27]

> Hence it is thus possible for the movement of the heart itself to be determined and altered through the increased or decreased pressure of the blood from the periphery of the body. Thus with strenuous bodily activity, where the blood presses more strongly into the veins, and is also more strongly attracted by the lungs, with stronger use of the blood in the periphery, and thus increased need for respiration, . . . in consequence of the stronger flow of blood to the heart, the movement of the heart itself increases. In these circumstances it is not the heart whose pulsations originally increase . . . but rather both the peripheral vascular systems, which only later evoke its heightened activity (p. 319).

Here, it is evident how, in Romantic physiology with its new emphasis on the relation between cardiac action and the flow of blood, there arose the increased attention to the circulatory and metabolic regulation of the periphery and its consequent effect on the heart that had been so significant in the Galenic system. This called for the investigation of these phenomena in the years that followed, and the same interest is still influential today in the conception of the function of the heart and circulation.

Carl Gustav Carus (1779-1868), in his *System der Physiologie* (1839), once more summarizes essential elements of "Romantic" physiology. His conception of the function of heart and circulation is determined above all by embryology: the formation of the heart explains its movement. According to Carus, islands of blood with connecting canals first form from the embryonic fluid; in these, the blood begins to circulate, affected by the electric polarity of the two cotelydons "as if through the poles of an electric apparatus."[28]

> He who fails to open his eyes, in order to grasp the extraordinarily remarkable process of the circular flow of the blood in its embryonic form, will understand it still less in a state of developed organization, for all his appeal to experiments and to analogies with pumps and pressure mechanisms (p. 39).

At the place where the various circular paths intersect, there arises through the pressure and the electrical polarity of the inversely moving fluids a kind of area of stimulation, in which the sensitive wall of the vessel begins to twitch and forms the *punctum saliens*.[29] But the twitching interrupts the stream that evoked it: without its stimulation, the cardiac Anlage relaxes and leaves the path free again. This "reciprocity" of vessel and content (p. 47) continues throughout the whole life of the organism.

In this process, the heart represents only the "mechanism in the life of the circulation"; "mechanism does not encompass life, but life includes mechanism in itself" (p. 83, 86). The movement of the blood remains the primary movement, even though "with currents that become ever broader, larger and more frequent ... the mechanical power of the stronger pulsation of an ever growing musculature [i.e. of the heart] to accelerate and maintain this flow becomes a second chief impelling force" (p. 87).

Evidence that "it is not cardiac movement alone that is the lever of the movement of the blood" is also found, according to Carus, in the slow centripetal flow of blood after death or after the cessation of the heartbeat; the blood then collects in the venous system (p. 146). But this, the "night side of the vascular system," corresponds to the blood of the early embryonic period, when the blood has not yet been exposed to the light. Thus, the blood tends in the end to move "toward the side, at which . . . the life of the blood has its very beginning, so that in this way too [it] . . . completes a true circular course. It is only and alone this general necessity of every living development . . . to incline again at its end to its beginning . . . that is also the ground . . . for the cessation of such a phenomenon" (p. 146). Thus, Carus also takes up again the idea of the *primum vivens ultimum moriens* (the first to live is the last to die) and relates it to the blood. This can serve as the symbolic summation of the "Romantic" conceptions of the function of heart and circulation.

Looking back, we find that in the physiologists of the second half of the eighteenth century and of the Romantic movement, aspects of Harvey's views recur: the epigenetic approach to embryology and the understanding of function on the basis of morphogenesis; the primacy of the blood over the heart, which is first formed and consolidated from the movement of the blood, and acts in reaction to that movement; the derivation of the mechanical aspect from the vital for the sake of expansion and "amplication"; and finally, the idea of polarity, which was not developed philosophically by Harvey but was nevertheless present implicitly in his thought. Indeed, even the reliance on analogy comes to life again in Romanticism: for example, in Oesterreicher's "astronomical" reading of the relation between the blood and the nervous system, or in Schultz, who for the first time tries to apply consistently to the reality of the organism the image of circumnavigation.[30]

Granted, there are also differences. Above all, Harvey sees the circulation from a unifying as well as centralizing point of view: the heart rules the circulation, whose "gradient" is produced only by the distance of the blood from its point of origin; not the periphery, but the center is the primary, the "more vital," and hence, as it were, the more "valuable" factor. In this picture, the doctrine of the central vital heat still holds, and also the Aristotelian preference for the heart as the chief organ of the soul.

Yet, as we have seen, Harvey himself already takes new paths: the notion of an elementary sensibility of the tissues, the explanation of cardiac activity on the basis of the polarity of the movement of the blood, but above all, in his case, along with the primacy of the blood as the ubiquitous life stuff of the organism—a conception that becomes more and more prominent in his work. All this points in the direction of a "systemic" vitalism, which is already reflected in Harvey in a changed localization of the regulation of the circulation, namely from the center to the periphery. These tendencies of Harvey's prefigure the beginning of a *countercurrent* to the predominant mechanistic thinking, which at first remains in the background—although, at the same time, it shines through even in the mechanistic theories, for instance, of cardiac action—but becomes increasingly dominant from the middle of the eighteenth century. To be sure, this happens without any awareness of their connection with Harvey on the part of the proponents of the new position; he is seen wholly against the background of his mechanistic reception.

On the other hand, it did not go very differently for the person who had been just as essentially of influence here. For the most part, the effect of Descartes's work must be inferred indirectly, even for the seventeenth century. With the ascendance and victory of the mechanistic-physicalistic paradigm from about 1830 on, as plainly as it carried on the Cartesian thought-style, Descartes's influence can no longer be attested in detail.[31] If they made any historical references, the protagonists of this paradigm invoked Harvey, who, unlike Descartes with his speculative theories, could be represented as a sober experimental scientist and could even be brought into the field against the vitalists. Thus, *François Magendie* (1783-1855) uses Harvey to attack the "reactionary" theories of the circulation based on peripheral flow:

> It is natural to assume that the pulmonary pump [the right heart—auth.], after it has driven the blood into the main duct, continues to move it in its large and small branches . . . and thus also in the pulmonary veins. This was *Harvey's* opinion, although that of the more recent physiologists is different. . . . Harvey discovers the true circulation of the blood, and immediately every one screams at him from all sides. . . . However, truth finally won out; Harvey's triumph seemed to be assured forever. And yet it was reserved for our time to consider once more a question that the English physiologist had so happily answered. To whom does the honor belong, or rather the shame, of this backwards-striding revolution? *John Hunter* and the Montpellier school have contributed a good deal to it . . . (But *Bichat* has) contributed the most . . . to bring the theory of the circulation of the blood back into the darkness from which it had so happily emerged . . .[32]

Nevertheless, Magendie's basic point of view is clearly not Harvey's, but rather the *Cartesian*:

> I see in the lung a pair of bellows, in the trachea a wind tube, in the glottis a mouthpiece. And is not that hydraulic machine, which is arranged so as to circulate the blood in our tissues, a marvel of mechanics? One could understand how, even without laws of life, the circulation would have to persist in the cadaver, if we could set in motion the whole system of pumps and pipes that the heart, the veins and the arteries represent. So true is it that here physical phenomena play the greatest part.[33]

The blood neither possesses its own vital motive power (p. 268), nor the capillaries their own contractility; phenomena of "irregular" blood movement are to be ascribed to the elasticity of the containing vessels (p. 236ff.). The "apparatus of the circulation as a whole . . . shows us a hydraulic *central machine* (the heart), which is both pump and container for the fluid, and guides the latter to all points of the body, *pipes leading outward* (the arteries) . . . and *pipes leading back* (the veins)"; thus the rule of the heart is once more unlimited.

In Germany, the swing to the new paradigm begins somewhat later (about 1840-1850). *Johannes Müller* (1801-1858), as a student of Philipp Franz von Walther—himself close to *Naturphilosophie*—in his *Handbuch der Physiologie des Menschen* (1837), distances himself from the vitalistic interpretations of the circulation:

> If transparent parts with blood flowing through them are examined in broad daylight, . . . not the slightest trace of an autonomous movement of the single molecules of blood is observed in the blood vessels, neither an attraction and repulsion of the blood corpuscles, nor of the particles of the blood fluid.[34]

Müller discusses at length the phenomena of independent blood flow and ascribes them to the elasticity of the vessels and other external conditions; thus, for example, the movement of fluids in the lower animals described by Carus is attributed to oscillating cilia (p. 150f.):

> I therefore deny the unique power of propulsion and assume only the vital interaction . . . between substance and blood, through which, under otherwise uniform conditions, a more lively part takes up more blood than otherwise, or than other parts . . . (p. 151).

A longer section in Müller's *Handbuch* is also devoted to the various theories of cardiac action. Here, we must take another look backward. Although Haller's or Fontana's explanation was widely accepted in or about 1800, it was not without opposition. Particularly in association with the experimental physiology that was developing, first of all in France, attempts were made to revive the latent mechanistic tradition of central control by the heart. Thus, *César Legallois* (1770-1814) contested an autonomous vitality of the organs and localized the principle of movement of the heart, not in the brain, but in the spinal column and its ganglia.[35] From the other side, "Romantic" physiologists also criticized the mechanical-deterministic stimulus-response sequence and the "external movement" of the heart in Haller's theory. For example, Philipp Franz von Walther rejected it in his *Physiologie* in these words:

> It is simply not the case that the pulsating movement of the heart is in any way communicated to it mechanically from without or through the effect of stimulation; for the heart goes on beating under the bell, in empty space ... Thus *Galen* was justified in ascribing to the heart its own, independent *pulsative* force (*ibid.*, p. 31).

This power of rhythmical action, Walther tells us, is determined essentially by the electrical polarity of the right and left heart (p. 34). The heart is something "living independently for itself" (p. 11); it does indeed possess a special irritability independent of the nerves, which, however, remains to a certain extent *directed to itself*:

> The heart is the most irritable structure of any, the focus of the irritability that emerges most purely as such, and is subordinated to no higher function, the center of irritable life (p. 33).

Thus, the idea so significant for subsequent thinkers of a "self-stimulation" of the heart is developed for the first time in the vitalistic physiology of Romanticism—with an appeal to the Galenic "pulsative faculty."

Müller, who was Walther's student, also criticizes Haller's theory, but in such a way that he sketches in his reflections a compromise between the mechanistic and vitalistic explanations that will be competing in the succeeding period:

> As necessary ... as a certain quantity of blood and a certain filling of the cardiac cavities is for the maintenance of the activity of the heart, and as certain as it is that every mechanical expansion from within must call forth contraction in it, still it is not the stimulation of the blood ... that is the final ground of the rhythmic contractions ... For the heart emptied of blood still continues its contractions in a weaker form (p. 88).

Nor does the law of the refractory phase discovered by Fontana rescue this theory: an "environmental stimulus", for instance, from the air, explains neither cardiac action nor its special consequence, for

> ... the same thing happens in empty space, and without an inner reason the regular sequence of ventricular contraction following the contraction of the auricles could not be maintained. Thus the cause must lie much deeper (p. 188).

Nor, moreover, can the solution lie in the cardiac nerves, that is, in a central release. Müller now postulates a "specific influence ... of the nerves *yet remaining* in the substance of the excised, empty heart ..., which appears to be the ultimate cause of the contractions of the heart" (p. 191; author's italics). Thus, cardiac action would be, in Galenic terminology, an intermediate between "animal" and "natural" movements: regulated neither by the central nervous system nor by purely muscular irritability. Thus, despite a certain resignation Müller already comes very close to the later solution:

> There must ... be something in the interaction of the cardiac nerves and the substance of the heart, which is either continuously active, but to which the heart ... (according to Fontana's law; auth.) reacts only periodically, or which itself affects the heart periodically. The answer to this question is infinitely difficult, in the present state of knowledge impossible.[36]

In about 1850, the new ideas were already firmly established in German physiology; this can be illustrated in *Alfred Wilhelm Volkmann's* (1800- 1877) purely mathematico-physically constructed "Haemodynamics" of 1850. Volkmann sees "the most fundamental agreement between the processes of the movement of the blood and the phenomena that I have observed in dead mechanisms."

Remaining incongruencies do not rest on a principle "which raises itself with freedom above the compulsion of physical laws, but . . . on the deficiency of observation on the one hand and on the other, on the inadequacy of the formulae."[37] In accordance with this conception, when the circulation is considered in general, it is a question of:

> . . . the progression of a fluid through pipes, which return to themselves in circles, and of the execution of this movement in a determinate time. This task is of a purely mechanical nature, and nothing prevents our assuming that the cardiac pump as a mechanical instrument is adequate to solve it (p. 314f.).

Previously, "theoretical prejudices" had occasioned distrust of the power of the heart. True, the phenomena of blushing and turning pale, of tumefactions, inflammations, and so, were not to be explained on basis of the heartbeat (cf. Bichat); " only no complaint can be made against the heart itself on this basis, for the production of those phenomena is not at all its business, but that of special apparatuses" (p. 325).

Volkmann, as well, takes issue in detail with the observations of the vitalists on the "primacy of the periphery" (pp. 325-341). As far as embryogenesis is concerned, he does indeed concede its "epigenetic" course, but nevertheless questions its significance for the adult organism. Elementary vital functions do indeed appear "in the germ . . . even before the development of definite organs, for the origination of these is already an act of life"; still "in mature animals nothing of this kind can occur, in them *the functions are determined by the organs and appear simply as an activity of those organs*: (p. 327; author's italics).

Just as little can phylogenesis provide insight into the causes of the circulation in higher organisms:

> Just as . . . it would occur to no one to consider the uterus dispensable in the act of birth by the female, because there are animals that produce their young without a uterus and without labor, so one should take care not to deduce the dispensability of the heart in the vertebrates from the fact that in many invertebrates a movement of fluid does indeed come about without a heart beat (p. 326).

Neither does Volkmann deny the second group of phenomena, namely, blood flow with non-moving or absent heart; but he interprets them as the balancing of differences in pressure (on the basis of continuing cardiac impulse) or differences in tension (on the basis of differential retractive force of arteries and veins). Instances of flow that last longer or that go backward find their explanation in intravasal coagulations, which delay the balance of tension (p. 329ff.). The hypothesis of an "attractive force" of organic tissues is also useless, Volkmann proceeds to argue: " . . . if this were correct, the mercury of a haemodynamometer that is introduced into the peripheral end of a severed artery would have to sink below zero; instead it rises" (p. 334). Finally, to base the circulation on the idea of polarity, for instance, on the determination of the arterial blood to become venous and vice versa, means to assert "that the idea toward which the organic process tends is the very force that carries it out"; but in this way, "the because of the end is substituted for the because of causality" (p. 355). As the valid result of his investigations for the future, Volkmann announces:

> The heart is a pumping engine and as such possesses enough force to drive the mass of blood in the circulation through the whole vascular system . . . Forces that support the achievements of the cardiac pump in a noteworthy way are not demonstrable. Muscular movements . . . may accomplish something, but really only a little (p.141).

As far as cardiac action is concerned, Volkmann too turns again against the irritability theory: it explains neither the "synchronism" of right and left heart nor the descending course of auricular and ventricular contraction (p. 377). Moreover, it is equivalent to the circle of the "self-impulsion" of the heart:

> Haller himself has shown that the cause of the movement of the blood lies in the thrust of the heart, and now, conversely, the blood stream is supposed to induce the movements of the heart. One time the heart is supposed to be like a pump, which does the driving, and another time like the mill wheel that is driven! (p. 381).

Fontana's modification of the theory is also said to be unacceptable: how could the "exhaustion" of the heart, despite continuing stimulation, lead to a "recovery" (p. 376). The blood, Volkmann declares, is only the common condition of *contractility*, not of contraction itself (p. 375). Instead, rather like Johannes Müller, he postulates an ordering "nervous central organ," *localized in the heart*, which, as all experiments to date have shown, is influenced by the cardiac nerves only in the sense of a frequency of modulation, but works primarily *independently* (p. 377f.).

As we see here, even the mechanistic tradition of an external control of the heart by the central nervous system finally ends with "self-stimulation" or *autorhythmy*—even if this proceeds from the *nervous* tissue peculiar to the heart. And so, in the end, Volkmann finds himself forced (if only in a footnote) to declare himself for an *autonomy of the organs and self-movement of the living*, for which the analogy of the self-movement of the heavenly bodies is not lacking:

> Fontana cancels entirely not only the independence of the movements, but also that of organic bodies as such, when he declares that every movement must have an external cause. A glance at cosmic relations suffices to show that an existence is possible which carries the conditions for its activity in itself. In the organism, which it is not without a deeper meaning to think of as a *microcosm*, similar relations recur. The greater independence that distinguishes living bodies from the non-living rests on the fact that they include a complex of conditions, which is *something mobile in itself* and which develops forces that determine one another reciprocally and without provocation from outside (p. 384; author's italics).

Here is an argumentation that would be worthy of a convinced vitalist in a mechanistic framework—this pretty paradox illustrates once more, in conclusion, the dialectical interaction of the two points of view in the development of today's interpretations of the heart and circulation. Two years later *Hermann Friedrich Stannius* (1808-1883), a student of Johannes Müller, demonstrated, through a ligature between the sinus of the vena cava and the auricle of a frog's heart, which resulted in cardiac inaction, the existence of an independent center of stimulation in the heart, the sinoatrial node.[38]

Notes to Part E

1. "Cor est principale et eminens machinae nostrae movens." Friedrich Hoffmann, *Fundamenta Medicinae ex principiis naturae mere mechanicis . . . proposita*, Halle, 1695, p. 43 (page references in what follows from this edition.)

2. Herman Boerhaave, *Institutiones Medicae, Opera omnia*, Venice, 1742, pp. 160, par. 415 (references below according to paragraphs).

3. François Boissier de Sauvages, *Dissertatio Medica de Motuum Vitalium*, 1741. In: Albrecht von Haller, *Disputationes Anatomicae*, Göttingen, 1746-51, vol. 4, p. 483. Cf. the excellent article by R.K. French,"Sauvages, Whytt and the motion of the heart: aspects of eighteenth century animism, *Clio Medica 7* (1972), pp. 35-54.

4. "A single soul, called rational according to its most important criterion, produces these organic activities immediately through its guidance and movement, without the mediation of spirits. "G.E. Stahl, *Dissertatio Physiologica-Medica de Sanguificatione in Corpore semel formato*, Halle, 1704, p. 10; quoted from French, p. 39.

5. In the *Dissertatio inauguralis medica de medicina medicinae curiosae* (1714), Stahl speaks of the soul as the principle "that stimulates and drives . . . the individual parts, which *are gifted with mechanical arrangement and mobility*, to the corresponding movements . . . " (quoted by Rothschuh, 1968, p. 154; author's ital.). Many "Stahlians" are not disturbed by allowing both levels to exist alongside one another and describing the actions of the soul purely mechanically. The unconsciousness of the vital functions is then explained (as in Borelli) as a *habit*, and this, in its turn, as the "facilitation" of the nerves through the continual flow of spirits (see French, pp. 40, 48). Descartes and the Cartesians had already introduced this principle as the explanation of memory, of association, and of learning (AT XI, p. 178, 184f.; p. 360, 368ff., p. 404ff.). The kinship between the two points of view is especially clear in the person of Boissier de Sauvages: at first a pure iatromechanist, he changed from about 1740 to being an animist who, nevertheless, continued to consider the body a "machine" (see Rothschuh, 1968, p. 157). All this shows that the frequent limitation of Descartes's influence to iatromechanism in the narrower sense does not include all of Descartes's real sphere of influence.

6. Hales (1677-1761) determined, among other items, blood pressure, volume of heartbeat, and vascular resistance; Bernoulli (1700-1782) and his student Daniel Passavant calculated the work of the heart among other things. This took place on a purely physical

basis: " . . . the animal Fluids move by Hydraulic and Hydrostatic laws" (S. Hales, *Statical Essays*, London, 1740, vol. 2, preface, p. 17); "the heart can be viewed as an hydraulic machine . . . " (D. Passavant, *De vi cordis*, Basel, 1748; quoted by Rothschuh, 1968, p. 121).

7. Johann Georg Zimmermann, *Dissertatio physiologica de Irritabilitate*, Göttingen, 1751, p. 67; similarly, p. 53.

8. " . . . ergo idem (cor), ubi sanguinem accepit, vi irritabili, et stimulo, quo fibrae in contractionem aguntur, constringitur, sanguine se evacuat, idemque stimulo liberum quiescit atque relaxatur." Albrecht von Haller, *Primae lineae Physiologiae*, Göttingen, 1751, p. 67; similarly, p. 53.

9. *Ibid.*, pp. 67f., 56. After decapitation of frogs, Haller found no immediate change in the movement of the heart; cf. A.v. Haller, *Opera minora*, vol. I ("Anatomica"), p. 233.

10. A.v. Haller, *Elementa Physiologiae corporis humani*, Lausanne, 1757-1766, vol. 4 (1762), p. 515f.

11. Cf. A.W. Volkmann, *Die Hämodynamik nach Versuchen*, Leipzig, 1850, pp. 370ff.; Rothschuh, 1968, p. 145f.

12. *Opera minora*, vol. I, p. 236f.

13. " . . . (sanguis) ergo videtur ad membranas corporis humani adtrahi . . . et in eas lineas congestum non solum stare, verum manifesto motu ad easdem . . . accedere . . . Porro semper mihi visum est, globulos ad alios globulos adtrahi, et ubicunque sanguis collectus fuerit, in trunco majoris arteriae, eo venire . . . nihil mihi probabilius videtur, quam sanguinem congregari, quacunque primo major eius moles sederit . . . " (*ibid.*, p. 238f.)

14. C.F. Wolff, *Theoria Generationis*, Halle, 1759, German, Berlin, 1764; reprint Hildesheim, 1966, p. 168f. (references below from this edition.)

15. Among others, *Heinrich Christian von Pander* (1794-1865) later made similar observations on the primary blood stream in his *Beiträge zur Entwicklungsgeschichte des Hühnchens im Eye*, Würzburg, 1817, pp. 14ff. On the other hand, *Karl Ernst von Baer* (1792-1876) expressed doubts about the matter in his famous work *Über die Entwicklungsgeschichte der Tiere. Beobachtungen und Reflexionen*, Königsberg, 1828-1837, vol. I, p. 31f.: "it seemed to me that movement is found first in the heart, somewhat later a flow in the grooves of the fruit and only after an additional flow of the red blood from the vascular district . . . It is not without hesitation that I present this account as the result of my investigations so far, since they have not corresponded to my conjectures. For it seemed probable, instead, that the heart was first supplied with blood through flow from the cotyledon, and so I would like to call for repeated investigations, for the formation of the

blood in warm blooded animals is subject to almost infinite difficulties..."

16. J. Hunter, *A Treatise on the Blood, Inflammation and Gun'Shot Wounds*, London (posthumously published), 1828. Among other things, Hunter says: "While the blood is circulating, it ... has the power of preserving its fluidity ...; or, in other words, the living principle in the body has the power of preserving it in this state" (p. 107);: "body, blood, and motion ... These three make up a complete body out of which arises a principle of self-motion" (p. 109); "this living principle in the blood... owes its existence to ... the materia vitae diffusa, of which every part of an animal has its portion" (p. 112f.).

17. K.F. von Kielmeyer,"Über die Verhältnisse der organischen Kräfte unter einander in der Reihe der verschiedenen Organisationen" (Stuttgart, 1793), reprinted in *Sudh. Arch. 23*, 1930, pp. 247-267, esp. p. 251. Animation of the blood is defended later, for example, by *Friedrich Ludwig Augustin* (1776-1854), *Lehrbuch der Physiologie des Menschen*, Berlin, 1809, vol. I, p. 327, or *Gottfried Reinhold Treviranus* (1776-1837), *Biologie oder Philosophie der lebenden Natur*, Göttingen, 1802-1822, vol. IV, p. 272f. On the concept of propulsive force, see also, Johannes Müller, *Handbuch der Physiologie des Menschen*, Koblenz, 1837, p. 149f.

18. F. X. Bichat, *Anatomie Générale*, Paris, 1801, pp.327ff., p. 509f. (Page references below according to this edition).

19. "Cessons donc de considérer cet organ comme l'agent unique qui préside au mouvement des gros vaissaux et à celuy de petits, qui, dans ces derniers ... produit l'inflammation, qui par son impulsion cause les diverse éruptions cutanées, les secrétions, les exhalations, etc. Toute la doctrine des mécaniciens reposoit, comme on sait, sur cette extrême étendue qu'ils avoient donnée au coeur pour ses mouvements" (p. 511).

20. P.F. v. Walther, *Physiologie des Menschen mit durchgängiger Rücksicht auf die comparative Physiologie der Tiere*, Landshut, 1807-8, vol. 2, p. 3f.

21. L. Oken, *Lehrbuch der Naturphilosophie* (1809-11), 2d. ed., Jena, 1831, p. 347f. *Johann Bernhard Wilbrand* (1779-1846) similarly applies the principle of polarity to the process of the circulation: "The movement of the blood, on the one hand to the heart, and on the other from the heart to all the structures of the body... these two movements form... with one another a true contradiction; they are related in entirely the same way as the two directions in the expression of magnetism, ... in the same way as positive and negative electrical charges. Without this inner contradiction in the whole movement of the blood ... the movement would not even be conceiv-

able . . . The polarity in the blood consists in the fact that the blood behaves chiefly materially in the system of the vena cava, whereas in the system of the pulmonary vein it has become animated by the act of respiration" (*Physiologie des Menschen*, Giessen, 1815, p. 127f.).

22. J.C. Reil, *Physiologische Schriften*, Vienna, 1811, vol. I, p. 216. In similar vein also I. Doellinger, *Grundriss der Naturlehre des menschlichen Organismus*, Bamberg, 1895, p. 210.

23. J. H. Oesterreicher, *Versuch einer Darstellung der Lehre vom Kreislauf des Blutes*, Nürnberg, 1826, p. 151f. (Page references below according to this edition).

24. P. 192, p. 196. Karl Heinrich Baumgaertner (1798-1886) develops a similar theory in his *Beobachtungen über die Nerven und das Blut*, Freiburg, 1830: the nerves exercise an "attractive" and a "repellent force" on the blood corpuscles with the former predominating (pp. 160ff.).

25. Carl Heinrich Schulz, *Das System der Circulation*, Stuttgart, 1838, p. 254.

26. Schultz refers to observations by the Viennese anatomist *Joseph Julius Czermak* (1799-1851), especially in the hibernating proteus; see *Medizinisches Jahrbuch des österreichischen Staates 15/1830:* p. 284.

27. p. 298f., p. 324. Oken expresses the same thought even more "galvanically": " The capillaries (of the body, auth.) thus attract the blood of the lung, separate it out, and form new components; and then, after it has acquired the same electricity, they push it back against the lung. The circulation exists only through the polarity between lung and capillaries . . . (*ibid.*, p. 346f.).

28. Carl Gustav Carus, *System der Physiologie*, Dresden, Leipzig, 1839, vol. II, p. 39 (Page references below to this edition). Earlier, Carus had already investigated the forms of circulation of fluids at different levels of the plant and animal kingdoms, and observed the simplest circular movements without a driving organ: " . . . thus the phenomenon of the circulation appears on three levels without a distinctive heart, and on three levels with more or less perfect hearts"; C. G. Carus, *Entdeckung eines vom Herzen aus beschleunigten Kreislaufs in den Larven netzflüglicher Insekten*, Leipzig, 1827, p. 30f.

29. P. 46f. According to Carus, without such an intersection only primitive forerunners of the heart can occur such as are found in lower organisms.

30. Oesterreicher justifies his analogy as follows: "If we draw a parallel . . . between the nervous system and the sun, but between the blood and the planets, then this is nothing more than a comparison, made in order to show that the laws according to which the universe exists, are everywhere the same—in the large as in the small, in

the macrocosm as in the microcosm; it is not a case of celestial laws applied to the animal organism, but of the laws of the universe exhibited both in the planetary system and in the animal organism" (*ibid.*, p. 192f.). *Die elliptische Blutbahn* (1809) by *Georg Ernst Vend* (1781-1831) offers an extreme example of analogizing speculation; here, we are told, in a parallel to Kepler's laws of planetary motion: "The squares of the movement of the blood or the time it takes the blood to circulate in its elliptical path are equal to the cubes of the mean distances from the bases of the blood's movement" (p. 63).

31. Nevertheless, according to T.S.Hall, the Cartesian tradition is really being continued here: "The soulless physiology practised by some in the days of Descartes, and explicitly championed by him, was thus delayed, but eventually it was decisive" (Commentary on the *Treatise of Man*, p. 114).

32. François Magendie, *Phénomènes Physiques de la Vie*, Paris, 1839, pp. 75ff.

33. P. 107.

34. Johannes Müller, *Handbuch der Physiologie des Menschen*, Koblenz, 1837, vol. I, p. 149.

35. Legallois carried out vivisection with decapitation and decreasing disturbance of the spinal cord (described in the *Biographisches Lexikon hervorragender Ärzte*, ed. A. Hirsch, Berlin, 1931, vol. III, p. 723, as "in part brutal and terrible"), in order to demonstrate against the vitalists a principle in the central nervous system for the control of the heart: "The heart receives all its powers from this same principle, the same as the other parts ... with the difference that the heart receives its powers from the whole of the spinal cord without exception, while each part of the body is animated only by a portion of that cord (namely the part from which it receives its nerves) ... We cannot admit this ... opinion of Bichat—although it has been generally adopted—that there exist in the same individual two different lives, animal life and organic life, that the brain is the unique center of animal life, and that the heart, independently of the brain and its nervous power, is the center of organic life." ("Le coeur emprunte toutes ces forces de ce même principe, de même que les autres parties ..., avec la différence que le coeur emprunte ces forces de tous les points de la moëlle sans exception, tandis que chaque partie du corps n'est animée que par une portion de cette moëlle (par celle dont elle reçoit ces nerfs) .... on ne peut plus admettre cette . .. opinion de Bichat, quoiqu'assez généralement adoptée, qu'il existe dans le même individu deux vies distinctes, la vie animale et la vie organique, que le cerveau est le centre unique de la vie animale, et que le coeur, indépendant du cerveau et de la puissance nerveuse, est le

centre de la vie organique." —César Legallois, *Expériences sur le Principe de la Vie*, Paris, 1812, p. 149, p. 152.

36. Müller, loc. cit. p. 118f.

37. Alfred Wilhelm Volkmann, *Die Hämodynamik nach Versuchen*, Leipzig, 1850, Preface, pp. V and VI.

38. Hermann Friedrich Stannius, "Zwei Reihen physiologischer Versuche," *Müllers Archiv*, Berlin, 1852, pp. 85-100; cf. Rothschuh, 1970, p. 84.

F

A Look Ahead

Summary. From 1850 to the present, the vitalistic aspect of the circulation remains present in latent form; this is evident in the occurrence of opposing and peripheral positions as well as in elements of the dominant thought style, which have their origin in the vitalistic tradition. In conclusion, this reciprocal relation between opposing points of view is considered in connection with the ambivalence of their object, especially the complex position of the heart in the organism: the historical sequence of changing perspectives also has its foundation in the actuality itself, which ever and again transcends a one-dimensional view and demands a different explanatory model.

The history of the interpretations of the heart and circulation after Harvey and Descartes as we have seen them in outline is itself clearly characterized by an "interplay" between the vital and the mechanical aspect, or between vital and mechanistic interpretations, which have not only replaced one another periodically, but have also reciprocally influenced and enriched one another, indeed were often characteristically fused with one another. Thus although the paradigm established about the middle of the nineteenth century still dominates physiology today in a modified form, elements of *both* traditions are nevertheless contained in the present view of the function of the heart and circulation.

An influence on the *mechanistic* tradition was already contained in Harvey's discovery itself, which, as we have seen, took place in an Aristotelian-vitalistic framework and nevertheless—extracted from its framework—served as the basis for the machine model of the body. But the conception of a sensibility and excitability of the muscular tissues and hence of an *autonomy* of the heart, which can be influenced by the central nervous system only in terms of modulation, also gained admittance to physiology under vitalistic premises with Harvey, Glisson, Haller, and the Romantic physiologists—and it still plays an essential role in the present conception. Finally, the attempt by Bichat and others to dethrone the "mechanical" center of the circulation directed attention once more to the vital processes and regulatory achievements of the periphery; and not least in importance, the rediscovery of the epigenetic approach and of comparative morphology in the latter part of the eighteenth century led to a better understanding of the developed heart and circulation.

But on the other side, the *vitalistic* tradition, too, is not without influence from its opponent. Thus, it was, above all, the atomistic interpretation of respiration and therewith the cancellation of the "vital heat" located in the heart that allowed the vitalists to take into account again the different quality of the two blood systems (in Romanticism their "polarity"), to apostrophize the blood in its transformations rather than the heart as the "carrier of life," and finally, in the twentieth century, to understand the metabolism of the organism as a steady state system in its relation to the environment. Further, it could be shown that it was precisely the Cartesian point of view, insofar as it banished from the body the soul that had formerly governed all the body's functions, that first opened a *space* that could now be used in a countermove for the conception of a pre-psychological vitality of the tissues and autonomy of the organs. After all, Haller was Boerhaave's student and not Stahl's.

Clearly, it is necessary not only to assume a reciprocal relation between dominance and latency, but also a reciprocal influence of different points of view on one another. Thus, a consideration of the dynamic of history could well contribute to shaking such points of view out of their seemingly incommensurable opposition and to bringing fixed positions once more into movement.

This seems all the more important as the time from 1850 to the present has been characterized by a lack of reflection on basic explanatory models in physiology. Vitalistic positions had only an outsider's role, which scarcely allowed a dialogue with the established paradigm that would be fruitful for both sides. The controversy about the priority of "heart" or "periphery" was silenced entirely in the second half of the nineteenth century; it was only in connection with the neovitalistic tendencies at the beginning of this century that the question once more arose, for instance, in the case of *Karl Hasebroek (Über den extrakardialen Kreislauf des Blutes,* 1914) or later, even more pointedly, in *Martin Mendelsohn (Das Herz, ein sekundäres Organ,* 1928).[1] If these attempts had little influence on a revitalization of the discussion, still, even among established physiologists, awareness persisted of problems that were unresolved, for instance, in connection with the reverse flow of the veins when sitting or the state of the blood stream in hearts with defective valves or the lymphatic circulation.[2] Even today attempts are again being made to question the "unique dominance" of the heart; the last word on this question has not yet been spoken.[3]

Moreover, the question of the cause and character of cardiac activity also continues to permit a variety of answers. The autorhythmy of the heart, which was demonstrated about the middle of the nineteenth century, corresponded neither to the theory of an external regulation of the heart by the central nervous system nor to the conception of a simple stimulus-response relation. Nevertheless, both explanatory traditions were to some extent justified. The influence of the *cardiac nerves*, not only on frequency, but also on contractility and on the guidance of excitation, is uncontested. "*Irritability*" or excitability as well as the refractory phase discovered by Fontana, that is, the phenomenon of *non-tetanizability* (in contrast to other muscles) are conditions of rhythmicity in the heart *muscle*. Finally, at the beginning of the 20th century, *Otto Frank* and *Ernest H. Starling* even showed an influence of the venous blood flow into the heart on its action (namely, via the expansion at the end of diastole). In this connection, Starling spoke of "a power of adaption" of the heart, "as suggestive of purpose as that of a sentient being," and called it "the most essentially 'vital' of the characteristics of the heart."[4]

As far as the nature of autorhythmy itself is concerned, much remains unexplained—indeed, even for the models of electrical sawtooth waves that are usually invoked, Magendie's dictum holds that in physiology "the explanation is only another expression for the fact."[5] It could well be said that the modern explanation of the Galenic "pulsative faculty" is not as far removed from the original as it appears. It is not surprising, therefore, that there are attempts to replace it by a "more mechanistic" theory. Thus in 1930, *Georg Hauffe* was of the opinion that the rhythmicity of cardiac movement could be attributed to purely non-organic phenomena of oscillation and flow although he did admittedly have to assume an embryonic blood stream independent of the heart.[6] Ironically, his attempt agrees with the vitalistic criticism of the conception of the heart as a pump—a criticism still heard today, which claims that the heart is rather an organ partly driven by the primary blood stream and serving to regulate it, a kind of "stream-transformer."[7] From this, there follows a surprising conclusion: *the mill-wheel conception of the heart*, of which Thomas Bartholinus first spoke, *can evidently be interpreted in mechanistic or vitalistic fashion—just as, conversely, the "pump" conception does indeed make the circulation an hydraulic mechanism, but at the price that it is referred to the vital autonomy, the "pulsative faculty," of the heart.*

Here, it seems, we hit on an essential motivation of the historical dynamic we have been considering: the change and the sometimes paradoxical intertwining of points of view in the physiology of the heart and circulation is not to be explained only by the general development in intellectual history, but is grounded in the dialectic of its object. Clearly, the polarity of the heart in itself and of heart and circulation as a whole is not automatically to be subsumed under one standpoint; every one-sided attempt to consider it seems to be unfair to essential aspects of the matter. We have met such polarities a number of times in the course of our investigation and will consider them once more, in conclusion, in the case of the heart itself.

By its very nature, cardiac activity stands between the voluntary ("animal") and the vegetative ("natural") movements that remain wholly in the unconscious; this difference is already expressed in the peculiarity of the cardiac musculature as distinct from the striated and the smooth muscles. The conduction system also takes an intermediate position insofar as it cannot be subsumed unambiguously either under the nervous or the muscular tissues, and thus, in Haller's terminology, unites in itself nervous "sensibility" and (muscular) "irritability."[8] In this context, the sinoatrial node is only the "most irritable" center in the heart, which is capable of producing excitation everywhere; it can be considered phylogenetically as well as ontogenetically as the "zone of stimulation" that differentiates itself and then makes itself independent at the *turning point* of the blood stream.[9]

The heart needs this relative independence precisely because it acts as *equilibriating organ* between various polar influences: above all, between the arterial and venous systems, between respiratory and metabolic functions, between sympatheticotonia and vagotonia, between the states of work and rest of the entire organism. The heart not only moves itself, it perceives these influences in the form of filling and resistance, in the expansion and tension of its wall, through its own receptors in the auricle and chamber, or through modulation by the central nervous system; it is just as much reacting to the blood stream as it is regulating and driving it. To understand the circulation only as a central motor, as is often done, is to be unjust to the polarity unique to it: precisely the investigation of pathological states (as in the case of shock) has showed this more and more clearly. Cardiac activity is exhausted neither by the "mill-wheel" nor by the "pump" model.

The manner in which the heart meets these complex influences, balances them, and unites them in itself is its polar self-movement, its rhythm—that through which it becomes "as it were like a living being on its own." The various influences from the whole organism are reflected in the cardiac rhythm and then communicate themselves to it like a reflection of its condition. The heart maintains the movement of the blood and, at the same time, gives it a new impulse as Harvey illustrated with the example of the person playing ball (see note C 104). And it appears that this very rhythmicity of the movement of the blood is of significance for the vitality of the tissues.[10] Still further, other rhythmic processes are oriented to the heartbeat, for example, breathing, which tends toward a whole number relation to the movement of the heart, or the pace, which (without conscious interference) harmonizes more or less with the pulse. Even the length of our musical sense of tempo corresponds, for example, with that of the pulse: a sequence of beats under 40/min we find musically incoherent, rhythms over 140/min, on the other hand, we feel to be hectic and disquieting. It is, perhaps also, to this capacity of the heart for the formation of rhythms that we can attribute the fact that, unlike any other organ, it is equipped to mirror feelings in time and space; for the psychological sphere is, after all, characterized precisely by polar states— joy and sorrow, anger and dread, sympathy and antipathy.

This multiplicity of relations and polarities, which has been just indicated here, is mirrored in the ambivalence in the effort to grasp the processes of the heart and circulation theoretically like the change of points of view in the historical development or like the contrast between Harvey and Descartes. It is not to be assumed that this interplay has found its end in the conceptions of today. Against this background, it is perhaps worth noting once more what Walter Pagel has written about Harvey: "unification of what is today sound and relevant with its apparent opposite," the attempt to tie together what appears self-contradictory, to allow a synchronic multiplicity of points of view—this could be a way to be fairer to the aspects of the heart and circulation. The present study is an attempt to facilitate the historical orientation necessary for such an understanding.

NOTES TO PART F

1. K. Hasebroek, *Über den extrakardialen Kreislauf des Blutes*, Jena, 1914, where, for example, we find the statement: "Apart from the minimal propulsive force that may still come from the arterial capillaries, the pressure

drop necessary for the return flow to the heart ... is also produced by *independent drives* of the walls of the veins directed toward the heart, which ... are at once aspiratory and propulsatory in the downward stream" (p. 177f.). M. Mendelsohn, "Das Herz—ein sekundäres Organ. Eine Kreislauftheorie", *Z. f. Kreislaufforschung 19* (1928), pp. 577-583. Mendelsohn writes: " . . . the movement of circulation has its beginning in the capillaries, just as everywhere else no stream begins at its mouth. The primary movement in the enclosed tube of the vessels then arises through the fact that considerable quantities of fluid are continually withdrawn from it and again reintroduced at other places. This motor power proceeds most obviously from the glands . . . " (p. 579f.) The heart is then primarily a regulator of the blood stream, "only in a secondary respect is it a motor" (p. 581).

    2. Cf. Pestel and Liebau, pp. 2ff., also, in the same volume, G. Liebau, "Über periphere Blutförderung" (pp. 67-78), where the idea of a "peripheral heart" is taken up again.

    3. Today, the primacy of the periphery is again vehemently put foward from the side of anthroposophically oriented medicine; see e.g. F. Husemann and O. Wolff, *Das Bild des Menschen als Grundlage der Heilkunst*, 3d. ed., Stuttgart, 1986, vol. III, pp. 95fff., p. 144f.: "It is decisive that the primary flow does not occur mechanically as cause of the movement of the blood, and is therefore not to be understood through mechanistic considerations. It is effective all the way to the veins ... (The) mechanical output is necessarily added to the primary flow coming from the periphery. Even when this is lacking...the inverse flow takes place, although with some difficulty ... The same holds for the flow of lymph . . . ." Cf. also L. Manteuffel-Szoege, *Über die Bewegung des Blutes. Hämodynamische Untersuchungen*, Stuttgart, 1977. Through comprehensive investigations, which take up again many of the vitalistic lines of argument, Manteuffel-Szoege tries to demonstrate a "proper inherent energy of movement" of the blood (p. 25). Even the regulation of the peripheral circulation is said to proceed to a certain extent from the blood itself (as Harvey had already seen it to do): "The self-movement of the blood also decides ... on the dislocation of the blood ... Thus turning red, which happens as the a symptom of anger or shame, and also turning pale, which happens after fright, are not the effect of pure vaso-motoric forces, but these phenomena are caused by the self-movement of the blood ... so that the movement of the blood ... is connected (with the life of feeling)" (p. 77).

    4. " ... its power of adaption, as suggestive of purpose as that of a sentient being"; " ... the most essentially 'vital' of the characteristics of the heart is its power of adaption ... " E.H. Starling, Linacre Lecture on the Law of the Heart, London, 1918, reprinted in P. A. Chevalier, ed., *The Heart and Circulation*, Stroudsberg, PA, 1976, pp. 84-110; quotations from p. 85, p. 87.

    5. See note E 36. On autorhythmy, cf. E. Schütz, *Physiologie des Herzens*, Berlin, 1958, pp. 45ff.

    6. G. Hauffe, *Herz, Pulsation und Blutbewegung*, Munich, 1930. Hauffe criticizes the current theory of autorhythmy as "too vitalistic": "The repeti-

tion is already presupposed as a property already inherent in the living organism, as a "potency," as a living process. The cell is "gifted" with automatic movement. A special force is supposed to reside in it, which also takes care of the rhythm. That the rhythm is a simple mechanism is not considered. But in this way, *every attempt to resolve mechanically the rhythm in the living organ is cut off in advance*" (p. 15). Using the model of an elastic rubber hose, Hauffe then tries to explain pulsation as a necessary process to be found in inanimate nature as well. The dynamic of elastic oscillation and of hydraulic waves keeps the heart in motion. "There is no ... cardiac muscle gifted with rhythmic autonomy, ... ; on the contrary, rhythm means repetitive movement in the elastic, oscillating system ... " (p. 241). True, in the course of his exposition he has to return to the vitalistic notion of a primary blood flow and reinterpret it mechanistically (pp. 41ff.).

7. In the anthroposophical conception, the heart is frequently considered on the model of the "hydraulic ram," an apparatus that transforms the kinetic energy of water that is already flowing into potential energy by a sudden interruption of the flow; see Manteuffel-Szoege, pp. 36ff., p. 47: "Thus if we want to assume that the heart works like a hydraulic ram, then we have to assume that the movement of the heart ... depends of the strength of the blood flow ... The role of the cardiac muscle is only to regulate the flow of blood in its cardiac portion." "The ram ... receives its force from the water flowing through it. What is of fundamental significance is the propulsive power of the water" (p. 53). In this connection, Manteuffel-Szoege, like Hauffe, speaks of the "laws of hydraulics" in effect here (*loc.cit.*).

8. The question of the primacy of "center" or "periphery" in embryonic development cannot be discussed here. It is clear that, in fact, there exists an independent peripheral flow comparable to the movement of the sap in plants, a flow which nevertheless emerges as a *directed* circulation only in connection with the beating heart: the circulation of the blood arises from the *polarity* of the "fluid" and the "solid." Cf. also K. L. Moore, *Embryologie*, Stuttgart/ New York, 1985, p. 74f.

9. In cardiac physiology, the conflict raged for a long time between the "neurogenic" and the "myogenic" theory of the formation of excitation (cf. Schütz, p. 33f.); today, there is an inclination toward the latter view since the heartbeat appears embryologically before the immigration of ganglia and nerve cells into the region of the heart.

10. The *pulsative* perfusion of isolated organs clearly sustains their function and prolongs their time of survival as compared with continuous flow; see K.J. Paquet, "Hämodynamische und metabolische Studien an isoliert durchströmten Schweinenieren mit pulsatorischer und kontinuierlicher Perfusion". In: Pestel and Liebau, pp. 53-66.

BIBLIOGRAPHY

LITERATURE TO 1900

Aristotle. *Works.* 23 volumes. Edited by Immanuel Bekker. Berlin, 1831-1870.

Bartholinus, Thomas. *Anatomia Reformata.* Leyden, 1651.

Baumgaertner, Karl Heinrich. *Beobachtungen über die Nerven und das Blut in ihrem gesunden und krankhaften Zustande.* Freiburg, 1830.

Bichat, François Xavier. *Anatomie générale.* Paris, 1801.

Blancaard, Stephen. "Institutiones Medicinae." In *Opera Medica, Theoretica, Practica et Chirurgica.* Leyden, 1701.

Boerhaave, Herman. "Institutiones Medicae." In *Opera Omnia,* 1–160. Venice, 1742.

Bohn, Johannes. *Circulus Anatomico-Physiologus, seu Oeconomia Animalis.* Leipzig, 1686.

Bontekoe, Cornelius. *Metaphysica; De Motu; Oeconomia Animales.* Leyden, 1688.

Borelli, Giovanni. *De Motu Animalium.* Rome, 1680–81.

Carus, C.G. *Entdeckung eines einfachen vom Herzen aus beschleunigten Blutkreislaufs in den Larven netzflüglicher Insekten.* Leipzig, 1827.

———. *System der Physiologie.* Dresden, Leipzig, 1839.

Carleton, Walter. *Exercitationes Physico-Anatomicae de Oeconomia Animali . . . Mechanice Explicata.* Amsterdam, 1659.

Conring, Hermann. *De Sanguinis Generatione et Motu Naturali* (1643). Leyden, 1646.

Craanen, Theodor. *Oeconomia Animalis, item Generatio Hominis ex Legibus Mechanicis.* Amsterdam, 1703.

Descartes, René. *Oeuvres.* Edited by C. Adam and P. Tannery. Paris, 1971-1974.

———. *Philosophical Writings of Descartes.* 3 volumes. Edited by John Cottingham, Robert Stoothoff, Dugald Murdoch and Anthony Kenny. Cambridge, 1985-1991.

Doellinger, Ignaz. *Grundriß der Naturlehre des menschlichen Organismus.* Bamburg, 1805.

Ent, Giorgio. *Apologia pro circulatione sanguinis, qua respondetur Aemelio Parisano, medico Veneto.* London, 1641.

Galenus, Claudius. *Opera*. Edited by C. Kühn. Leipzig, 1821–33.
Glisson, Francis. *Tractatus de Ventriculo et Intestinis*. London, 1677.
Hales, Stephen. *Statical Essays*. London, 1740.
Haller, Albert von. *Disputationes Anatomicae*. Göttingen, 1746–51.
———. *Elementa Physiologiae Corporis Humani*. Lausanne, 1757–66.
———. *Opera Minora*. Lausanne, 1762.
———. *Primae Lineae Physiologiae*. Göttingen, 1751.
Harvey, William. *Opera, sive Exercitationes de Motu Cordis et Sanguinis in Animalibus atque Exercitationes duae Anatomicae de Circulatione Sanguinis tumque Exercitationes de Generatione Animalium: Quibus Praefationem addidit Bernardus Siegfried Albinus*. Leyden, 1737.
———. *The Works of William Harvey*. Edited by R. Willis. London, 1847.
———. *De Motu Locali Animalium*. (*Local Movement of Animals*.) Edited and translated by G. Whitteridge. Cambridge, 1959.
———. *Praelectiones Anatomiae Universalis*. (*The Anatomical Lectures of William Harvey*.) Edited by G. Whitteridge. Edinburgh, 1964.
Highmore, Nathaniel. *Corporis Humani Disquitio Anatomica*. The Hague, 1651.
Hippocrates. *Oeuvres complètes d'Hippocrate*. Edited by E. Littré. Paris, 1839–61.
Ho(o)g(h)elande, Cornelis van. *Cogitationes, quibus Dei Existentia: item Animae Spiritalitas, et Possibilis cum Corpore Unio, Demonstratur: Nec non, Brevis Historia Oeconomiae Corporis*. Leyden, 1676.
Hoffmann, Friedrich. *Fundamenta Medicinae ex Principiis Naturae Mere Mechanicis . . . Proposita*. Halle, 1695.
Hunter, John. *A Treatise on the Blood, Inflamation and Gun-Shot Wounds*. London, 1828.
Kielmeyer, Karl Friedrich von. *Über die Verhältnisse der organischen Kräfte unter einander in der Reihe der verschiedenen Organisationen*. Stuttgart, 1793. Reprint in *Sudh.Arch*. 23 (1930): 247–67.
Legallois, César. *Expériences sur le Principe de la Vie*. Paris, 1812.
Levison, Georg. *An Essay on the Blood*. London, 1776. German: *Versuch über das Blut*. Berlin, 1782.
Lower, Richard. *Tractatus de Corde*. London, 1669.
Mayow, John. *Tractatus Quinque Medio-Physici*. Oxford, 1674.
Müller, Johannes. *Handbuch der Physiologie des Menschen*. Coblenz,, 1837.
Oesterreicher, Johann Heinrich. *Versuch einer Darstellung der Lehre vom Kreislauf des Blutes*. Nuremberg, 1826.
Oken, Lorenz. *Lehrbuch der Naturphilosophie*, 1809–11. 2nd edition. Jena, 1831.

Pander, Heinrich Christian von. *Beiträge zur Entwickelungsgeschichte des Hünchens im Eye.* Würzburg, 1817.
Pascal, Blaise. *Oeuvres.* Paris, 1906.
Plemp, Vopiscus Fortunatus. *Fundamenta Medicinae.* Löwen, 1644.
Regius, Henricus. *Fundamenta Medica.* Utrecht, 1647.
———. *Fundamenta Physices.* Amsterdam, 1646.
Reil, Johann Christoph. *Physiologische Schriften.* Vienna, 1911.
Schultz (-Schultzenstein), Carl Heinrich. *Das Systeme der Circulation.* Stuttgart, 1836.
Steno, Nicolaus. *Opera Philosophica.* Copenhagen, 1910.
Sylvius, Franciscus. "Disputationes Medicae" (1659–63). In Franciscus de la Boe (Sylvius). *Opera Medica.* Paris, 1679.
Vend, Georg Ernst. *Die elliptische Blutbahn.* Würzburg, 1809.
Volkmann, Alfred Wilhelm. *Die Hämodynamik nach Versuchen.* Leipzig, 1850.
Wale, Jan de (Waleus, Johannes). "Epistolae duae: De Motu Chylis et Sanguinis ad Thomam Bartholinum." In: Caspar Bartholin. *Institutiones Anatomicae.* Leyden, 1641; Thomas Bartholinus. *Anatomia Reformata.* Leyden, 1651, 529–76.
Walther, Philipp Franz von. *Physiologie des Menschen mit durchgängiger Rücksicht auf die comparative Physiologie der Thiere.* Landshut, 1807–8.
Wilbrand, Johann Bernhard. *Physiologie des Menschen.* Giessen, 1815.
Willis, Thomas. *Cerebri Anatome.* London, 1664.
———. *Diatribe duae Medico-Philosophicae: De Fermentatione; De Febribus* (1659). 4th edition. London, 1677.
Wolff, Caspar Friedrich. *Theoria Generationis. (Theorie von der Generation.)* Halle, 1759; Berlin, 1764. Reprint Hildesheim, 1966.

B. LITERATURE SINCE 1900

Ackernecht, E. *Kurze Geschichte der Medizin.* Stuttgart, 1959.
Ballauf, T. *Die Wissenschaft vom Leben: Eine Geschichte der Biologie.* Freiburg, 1954.
Basalla, G. "William Harvey and the Heart as a Pump." *Bull. Hist. Med.* 26 (1962): 467–70.
Bayon, H.B. "The Lifework of William Harvey and Modern Medical Progress." *Proc.Roy.Soc.Med.* 44 (1951): 213–18.
Birkenhead, Lord C. of. "The Germ of an Idea or What Put Harvey on the Scent?" *J.Hist.Med.* 12 (1957): 102–5.

Bitbol-Hespériès, A. *Le principe de vie chez Descartes*. Paris, 1990.
Boas, Marie. "The Establishment of the Mechanical Philosophy." *Osiris*. 10 (1952): 412–541.
Böhm, W. "John Mayow and Descartes." *Sudh.Arch.* 46 (1962): 45–68.
Brown, T. "The College of Physicians and the Acceptance of Iatromechanism in England." *Bull.Hist.Med.* 44 (1970): 12–30.
Brunn, W. *Die Kreislauffunktion in William Harvey's Schriften*. Berlin, 1967.
Burchell, H.B. "Mechanical and Hydraulic Analogies in Harvey's Discovery of the Circulation." *J.Hist.Med.* 26 (1981): 260–77.
Butterfield, H. *The Origins of Modern Science, 1300–1600*. New York, 1957.
Bylebyl, J. "The Medical Side of Harvey's Discovery: The Normal and the Abnormal." In *William Harvey and His Age*, edited by H. Bylebyl. Baltimore, 1979.
Canquilhem, G. *La formation du concept de réflexe aux XVIIe et XVIIIe siécles*. Paris, 1977.
Carter, R.C. *Descartes' Medical Philosophy: The Organic Solution to the Mind-Body Problem*. Baltimore, 1983.
Castiglioni, A. *Histoire de la médecine*. Paris, 1931.
Clarke, D. M. *Descartes's Philosophy of Science*. Manchester, 1982.
Crombie, A. *Von Augustinus bis Galelei: Die Emanzipation der Naturwissenschaften*. Berlin, 1959; Munich, 1970.
Debus, A.G. "Harvey and Fludd: The Irrational Factor in the Rational in the Science of the Seveneenth Century." *J.Hist.Biol.* 3 (1970): 81–105.
Dijksterhuis, E. *Descartes et le cartésianisme hollandais: Études et documents*. Paris, 1951.
———. *Die Mechanisierung des Weltbildes*. Berlin, 1956.
Dreyfus-Le-Foyer, H. "Les conceptions médicales de Descartes." *Revue de Métaphysique et de Morale* 44 (1937): 237–86.
Driesch, H. *Geschichte des Vitalismus*. Leipzig: 1905, 1923.
Faller, A. "Niels Stensen und der Cartesianismus." In *Nicolaus Steno and his Indice*, edited by G. Scher, 140–66. Acta Historica Sc. Nat. et Med., vol. 15. Copenhagen, 1958.
Fischer, H. "Die Geschichte der Zeugungs- und Entwicklungstheorien im 17. Jahrhundert." *Gesnerus* 2 (1945): 49–80.
Fleck, Ludwig. Entstehung und Entwicklung einer wissenschaftlichen Tatsache (1935). Frankfurt, 1980.
———. *Erfahrung und Tatsache*. Frankfurt, 1983.
———. *Genesis and Development of a Scientific Fact*. Chicago, 1981.

Foster, M. *Lectures on the History of Physiology during the Sixteenth, Seventeenth and Eighteenth Centuries.* Cambridge, 1924.
Frank, R. *Harvey and the Oxford Physiologists.* Berkeley, 1980.
French, R.K. "Sauvages, Whytt, and the Motion of the Heart: Aspects of Eighteenth Century Animism." *Clio Med.* 7 (1972): 35–54.
Gasking, E. *Investigations into Generation, 1651–1828.* London, 1968.
Garrison, F. *An Introduction to the History of Medicine.* Philadelphia, 1929.
Georges-Berthier, A. "Le mécanisme cartésien et la physiologie au XVIIe siècle." *Isis* 2 (1914): 37–89; 3 (1920–21): 21–58.
Gilson, E. "Descartes, Harvey et la Scholastique." In *Études sur le rôle de la pensée médiévale dans la formation du système cartésien,* edited by E. Gilson, 51–100. Paris, 1984.
Goltz, D. "Der leere Uterus: Zum Einfluß von Harveys *De Generatione Animalium* auf die Lehren von der Konzeption." *Medizinhist. J.* 21 (1986): 242–68.
Grene, M. "Life, Disease and Death: A Metaphysical Viewpoint." In *Organism, Medicine and Metaphysics: Essays in Honour of Hans Jonas,* edited by S. Spicker, 233–63. Dordrecht, 1978.
Hall, A.R. "Studies on the History of the Cardiovascular System. I. Galen." *Bull.Hist.Med.* 34 (1960): 391–413.
Hall, T.S. "Descartes' Physiological Method: Position, Principles, Examples." *J.Hist.Biol.* 1 (1970): 53–79.
———, ed. René Descartes. *Treatise on Man.* Cambridge, Mass., 1972.
Hammacher, K. *Einleitung zu Descartes' "Leidenschaften der Seele".* Hamburg, 1984.
Hassbroek, K. *Über den extrakardialen Kreislauf des Blutes.* Jena, 1914.
Hauffe, G. *Herz, Pulsation and Blutbewegung.* Munich, 1930.
Hildebrand, G. "Arterielle Pulsation und rythmische Koordination." In *Phänomen der pulsierenden Strömung im Blutkreislauf aus technologischer, physiologischer und klinischer Sicht,* edited by E. Pestel and G. Liebau, 34–52. Mannheim, 1970.
Husemann, F., and O. Wolff. *Das Bild des Menschens als Grundlage der Heilkunst.* 3rd edition. Stuttgart, 1986.
Jacob, F. *Die Logik des Lebenden.* Frankfurt, 1971.
Jevons, F.R. "Harvey's Quantitative Method." *Bull.Hist.Med.* 36 (1962): 462–67.
Jonas, Hans. *Organismus und Freiheit.* Göttingen, 1973.
Kassler, J.C. "Man—A Musical Instrument: Models of Brain and Mental Functioning before the Computer." *Hist.Sci.* 22 (1984): 59–92.

Keele, K.D. *William Harvey: The Man, the Physician and the Scientist.* London, 1965.
Keynes, G. *The Life of William Harvey.* Oxford, 1966.
Kilgour, F.G. "William Harvey's Use of the Quantitative Method." *Yale J.of Biol.and Med.* 26 (1954): 410–21.
King, L. *The Growth of Medical Thought.* Chicago, 1963.
Kuhn, T.S. *The Structure of Scientific Revlutions.* Chicago, 1962.
Kutschmann, W. *Der Naturwissenschaftler und sein Körper.* Frankfurt, 1986.
Lamprecht, S.P. "The Rôle of Descartes in Seventeenth-Century England." *Stud.Hist.Ideas* 3 (1935): 182–242.
Lesky, E. "Harvey und Aristoteles." *Sudh.Arch.* 41 (1957): 289–318, 349–78.
Lindeboom, G.A. *Descartes and Medicine.* Amsterdam, 1978.
———. "The Impact of Descartes on Seventeenth Century Medical Thought in the Netherlands." *Janus* 58 (1973): 201–6.
———. "The Reception in Holland of Harvey's Theory of the Circulation of the Blood.: *Janus* 46 (1957): 183–200.
Lindroth, S. *Descartes in Uppsala.* Stockholm, 1964.
———. "Harvey, Descartes and Young Olaus Rudbeck." *J.Hist.Med.* 12 (1957): 209–19.
Löw, R. *Philosophie des Lebendigen.* Frankfurt, 1980.
Manteuffel-Szoege, L. *Über die Bewegung des Blutes: Hämodynamische Untersuchungen.* Stuttgart, 1977.
Mendelsohn, E. *Heat and Life: The Development of the Theory of Animal Heat.* Cambridge, Mass., 1964.
Mendelsohn, M. "Das Herz—ein sekundäres Organ: Eine Kreislauftheorie." *Zeitschrift für Kreislaufforschung* 19 (1928): 577–83.
Mesnard, P. "L'esprit de la physiologie cartésienne." *Archives de Philosophie* 13 (1937): 181–220.
Mowry, B. "From Galen's Theory to William Harvey's Theory: A Case Study in the Rationality of Scientific Theory Change." *Stud.Hist.Phil.Sci.* 16 (1985): 49–82.
Pagel, W. "Harvey and Glisson on Irritability: With a Note on Van Helmont." *Bull.Hist.Med.* 41 (1967): 497–514.
———. "Harvey's Rôle in the History of Medicine." *Bull.Hist.Med.* 24 (1950): 70–73.
———. *New Light on William Harvey.* Basel, 1976.
———. "The Reaction to Aristotle in 17th Century Biological Thought." In *Science, Medicine and History*, edited by E. Underwood, 489–509.

Oxford, 1953.

———. *William Harvey's Biological Ideas.* Basel, 1967.

Passmore, J.A. "William Harvey *and the Philosophy of Science.*" *Austr.J.Phil.* 36 (1958): 35–108.

Pavlov, I. *Vorlesungen über die Arbeit der Großhirnhemisphären* (1927). In his *Sämtliche Werke*, vol. 4. Berlin, 1953.

———. *Psychopathology and Psychiatry.* Moscow, 1960.

Pellegrin, P. "Aristotle: A Zoology without Species." In *Aristotle on Nature and Living Things*, edited by A. Gohhelf, 95-115. Pittsburgh, Penn., 1985.

Peller, S. "Harvey's and Cesalpino's Rôle in the History of Medicine." *Bull.Hist.Med.* 23 (1949): 23–35.

Pestel, E., and G. Liebau, eds. *Phänomen der pulsierenden Strömung im Blutkreislauf aus technologischer, physiologischer und klinischer Sicht.* Mannheim, 1970.

Plochmann, G.K. "William Harvey and His Methods." *Studies in the Renaissance* 10 (1963): 192–210.

Popper, K., and J. Eccles. *The Self and Its Brain*, New York, 1983.

Rather, L.J. "Old and New Views of Emotion and Bodily Changes: Wright and Harvey versus Descartes, James and Cannon." *Clio Med.* 1 (1965): 1–25.

Rosenfield, L. *From Beast-Machine to Man-Machine: The Theme of Animal Soul in French Letters from Descartes to La Mettrie.* New York, 1940.

Rothschuh, K.E. *Geschichte der Physiologie.* Berlin, 1953.

———. "Geschichtliches zur Lehre von der Automatie, Unterhaltung und Regelung der Herztätigkeit." *Gesnerus* 27 (1970): 66–86. (= Rothschuh 1970)

———. "Henricus Regius und Descartes." *Arch.Int.Hist.Sci.* 21 (1968): 39–66.

———. *Konzepte der Medizin in Vergangenheit und Gegenwart.* Stuttgart, 1978.

———. *Physiologie: Der Wandel ihrer Konzepte, Probleme und Methoden vom 16.–19. Jahrhundert.* Freiburg, Munich, 1968. (= Rothschuh 1968)

———. *Physiologie im Werden.* Stuttgart, 1969. Including, among others: "Das System von Jean Fernel (1542) und seine Wurlzon," 59–65; Die Entwicklung der Kreislauflehre im Anschluß an William Harvey," 66–86; "Vom Spiritus animalis zum Nervenaktionsstrom," 111–38.

———. "René Descartes und die Theorie der Lebenserscheinungen." *Sudh.Arch.* 50 (1966): 25–42.

———. "Technomorphes Lebensmodell contra Virtus-Modell (Descartes gegen Fernel)." *Sudh.Arch.* 54 (1970): 337–54.

Rüsche, F. *Blut, Leben und Seele; Ihr Verhältnis nach Auffassung der griechischen und hellenistischen Antike, der Bible und der alten alexandrinischen Theologen.* Paderborn, 1930.

Schimank, H. "Der Aspekt der Naturgesetzlichkeit im Wandel der Zeiten." In *Das Problem de Gesetzlichkeit,* edited by the Jungius-Gesellschaft der Wissenschaften, 139–86. Hamburg, 1949.

Schütz, E. *Physiologie des Herzens.* Berlin, 1958.

Singer, C., and E. Underwood. *A Short History of Medicine.* Oxford, 1962.

Sloan, P.R. "Descartes, the Sceptics, and the Rejection of Vitalism in 17th Century Physiology." *Stud.Hist.Phil.* 8 (1977): 1–28.

Stannard, J. "Aristotelian Influences and References in Harvey's *De Motu Locali Animalium.*" In *Studies in Philosophy and in the History of Science: Essays in Honor of Max Fisch,* 122–31. Lawrence, Kans., 1970.

Starling, H. *Linacre Lecture on the Law of the Heart.* London, 1918. Reprint in *The Heart and the Circulation,* edited by P.A. Chevalier, 84–110. Stroudsburg, Pa., 1976.

Temkin, O. "The Classical Roots of Glisson's Doctrine of Irritability." *Bull.Hist.Med.* 38 (1964): 297–328.

———. "Metaphors of Human Biology." In *Science and Civilization,* edited by R.C. Stauffer, 167–94. Madison, Wisc., 1949.

———. "On Galen's Pneumatology." *Gesnerus* 8 (1951): 180–89.

Toellner, R. "The Controversy between Descartes and Harvey Regarding the Nature of Cardiac Medicine." In *Science, Medicine and Society in the Renaissance,* edited by A. Debus, 73–89. New York, 1972.

Unschuld, P.U. "Gedanken zu kognitiven Ästhetik Europas und Ostasiens." In *Kulturvergleich in Forschung und Anwendung.* Vortragsreihe der Akademie der Wissenschaften zu Berlin. Berlin, 1989.

———. *Medizin in China: Eine Ideengeschichte.* Munich, 1980.

Webster, C. "Harvey's *De Generatione*: Its Origins and Relevance to the Theory of Circulation." *Brit.J.Hist.Sci.* 3 (1967): 262–74.

Weil, E. "The Echo of Harvey's *De Motu Cordis* (1628) 1628 to 1657." *J.Hist.Med.* 12 (1957): 167–74.

Weinberg, K. "Zum Wandel des Sinnbezirks von 'Herz' und 'Instinkt' unter dem Einfluß Descartes'." *Arch.f.d.Stud.d.neueren Sprachen u.Literaturen* 203 (1966): 1–31.

Westfall, R.S. *The Construction of Modern Science.* New York, 1971.

Whitteridge, G. *William Harvey and the Circulation of the Blood*. London, 1971.

Zilsel, E. "Die Entstehung des Begriffs des physikalischen Gesetzes." In his *Die sozialen Ursprünge der neuzeitlichen Wissenschaft*, 66–97. Frankfurt, 1976.

Index of Names

A
Ackerknecht, E., 33, 93
Alquié, F., 175
Anaximenes, 22-23
Aristotle, x, xi, xiv, 15, 20, 23-6, 30, 35-7, 39-40, 52-4, 60, 64-6, 69, 71, 76-7, 81, 85-7, 93-4, 97-113, 117, 133, 139, 178
Aselli, G., 59, 152
Averroes, 110
Avicenna, 21

B
Back, J. de, 187
Bacon, Francis, 28, 33, 45, 93, 118
Baer, Karl Ernst von, 219
Ballauf, T., 33, 93
Bartholinus, C., 142, 194
Bartholinus, Thomas, 169-70, 187, 194, 227
Basalla, G., 99
Bathurst, Ralph, 192
Bauhin, C., 180
Baumgaertner, Karl Heinrich, 221
Bell, Charles, 181
Bernard, Claude, 1, 15
Bernoulli, Daniel, 200
Beverwijck, Johan van, 179, 185
Bichat, François Xavier, 197, 204, 206, 211, 215, 220, 222, 226
Birkenhead, Lord C. of, 97
Bitbol-Hespériès, A., 179-80
Blancaard, Stephen, 142, 156-7, 190
Boas, Marie, 184

Böhm, W., 180, 193-4
Boerhaave, Herman, 199-200, 218, 226
Bohn, Johannes, 142, 156-7, 171, 173-5, 186, 195
Bontekoe, Cornelis, 142, 154-6, 189-90
Borelli, Giovanni, 142, 145, 154, 156-7, 171-5, 186, 195, 218
Boyle, Robert, 45, 97, 158, 160, 182
Brown, T.M., 184
Brunn, W., 97-100, 106-8, 111
Bruno, Giordano, 45
Burchell, H.B., 99
Butterfield, H. 29
Bylebyl, J. 57, 97, 101

C
Canguilhem, G., 181
Carter, R.C., 182-3
Carus, Carl Gustav, 198, 209-10, 212, 221
Castiglioni, A., 33, 93
Cesalpino, Andrea, 55, 101
Charleton, Walter, 142, 170, 195
Clarke, D.M., 177
Cole, F.J., 113
Colombo, Realdo, 28, 44
Columbus, Christopherus, 6
Conring, Hermann, 142, 169, 194, 203
Craanen, Theodor, 142-3, 149, 153-4, 156, 189
Critias, 110
Crombie, Alistair, 16
Czermak, Joseph Julius, 221

D

Debus, A.G., 179, 185
Democritus, 82, 110
Descartes, René, xi-xiii, 1-7, 9-10, 12-3, 21, 30, 39, 115-4, 159, 161-2, 165-6, 172-93, 195, 198, 211, 218, 222, 225, 229
Diemer, A., 17
Digby, Kenelm, 159, 191
Dijksterhuis, E., 184
Diogenes of Apollonia, 23, 110
Doellinger, Ignaz, 205, 221
Dreyfus-Le-Foyer, H., 180
Driesch, H., 34, 93

E

Eccles, John, 179
Ent, George, xiii, 142-3, 158-9, 161, 191-2
Empedocles, 23, 82, 110
Erasistratos, 24, 194

F

Fabricius of Aquapendente, x, 44-5
Faller, A., 184
Fernel, Jean, 19-22, 25-31, 57, 109, 140
Fleck, Ludvik, x, xiii, 3-7, 16-7, 45, 97, 192
Fludd, Robert, 5, 7, 45, 185
Fontana, Felice, 202, 213-4, 216-7, 227
Foster, M., 188, 192-3
Frank, O., 227
Frank, R., 191-3
French, R.K., 218

G

Galenus, Claudius, 16, 20-31, 46, 57, 59, 66-7, 103, 123, 132, 136, 139, 157, 166, 168, 178, 181-2, 190, 213
Galilei, Galileo, 40, 96, 118, 171
Garrison, F., 5, 15-6, 93
Gasking, E., 107, 109, 113, 186
Gassendi, Pierre, 108, 177, 185
Georges-Berthier, A., 179
Gilson, Etienne, 31
Glisson, Francis, 63, 103-4, 142, 170-1, 173, 175, 190, 195, 200-1, 226
Goddard, Jonathan, 190
Goltz, D., 113
Grene, M., 178

H

Hales, Stephen, 200, 218
Hall, A.R., 31
Hall, T.S., 179, 181, 222
Haller, Albrecht von, 5-6, 14, 17, 104, 197, 200-3, 206, 213, 218-9, 226, 228
Hammacher, K., 181
Harvey, William, ix-xvii, 1-13, 19-20, 27-9, 31, 33-125, 128-9, 131-32, 133-4, 139-46, 149, 151, 156-9, 161, 166-71, 175, 179, 181-6, 188, 190-2, 194-5, 197-200, 202-4, 206-8, 210-1, 225-6, 229-30
Hasebroek, K., 227, 229
Hauffe, G., 227, 230-1
Hayman, 187
Helmont, Johann Baptista van, 104, 151
Heraclitus, 22-23
Herophilus, 24
Highmore, Nathaniel, 108, 142, 159-60, 191
Hildebrand, G., 190

Hippocrates, 30
Hoffmann, C., 102
Hoffmann, Friedrich, 149, 199, 218
Hoghelande, Cornelis van, xiii, 142-3, 148-50, 154, 161, 169, 184-5, 188, 194
Hooke, Robert, 158, 165, 182
Hunter, John, 203, 206, 211, 220
Husemann, F., 230

J

Jacob, F., 178
Jevons, F.R., 97
Jonas, Hans, 177-8

K

Kassler, J.C., 182
Keele, K.D., 95, 97-8, 185
Kepler, Johannes, 16, 118, 222
Keynes, G., 97
Kielmeyer, Karl Friedrich von, 203, 220
Kilgour, F.G., 97
King, L., 102
Kuhn, Thomas F., xiv, 3-10, 16-7, 29, 31, 39, 62, 95-6, 102

L

Lamprecht, S.P., 184, 191
Larkin, V., 178
Legallois, Çésar, 213, 222-3
Lesky, Erna, 33, 39, 93-5, 107, 113
Levison, Georg, 203
Lewinsohn (=Morus), 184
Liebau, G., 190, 230-1
Lindeboom, G.A., 179, 184-5, 187-8
Lindroth, S., 184-5
Löw, Reinhard, 17, 30

Lower, Richard, 142, 145, 154, 158, 160, 163-5, 174, 186, 193

M

Magendie, François, 181, 198, 211, 222
Malebranche, Nicolas, 186
Malphigi, Marcello, 154
Manteuffel-Szoege, L., 16, 230-1
Marci, Marcus, 109
Mayow, John, 142, 145, 156, 158, 160, 165-7, 180, 182, 186, 192-4
Mendelsohn, E., 30-1, 100, 182
Mendelsohn, M., 226, 230
Mersenne, Marin, 175, 177, 185
Mesnard, P., 179, 188
Moliére, Jean Baptiste, 31
Moore, K.L., 231
Morison, R., 95, 102, 105
Mowry, B., 5, 16
Müller, Johannes, 212-4, 217, 220, 222-23

N

Needham, J., 31, 34, 94, 107, 183
Newton, Isaac, 118

O

Ockham, William of, 27
Oesterreicher, Johann Heinrich, 16, 198, 206, 210, 221
Oken, Lorenz, 197, 205, 220-1

P

Pagel, William, xiv, 2, 15-6, 34, 42, 45, 93-4, 96-9, 101, 104, 109-10, 114, 195, 229
Pander, Heinrich Christian von, 219

Paquet, K.J., 231
Paracelsus, Theophrastus Bombastus, 98, 151
Pascal, Blaise, 183
Passmore, J.A., 185
Passavant, D., 218-9
Pavlov, Ivan, 181, 188
Pellegrin, P., 30
Pestel, E., 190, 230-1
Petty, William, 159, 191
Plato, 23-4, 30, 36
Plemp, Vopiscus Fortunatus, xii, 129, 179-80, 183, 186, 187-8
Plochmann, G.K., 34, 94-5
Popper, Karl, 179
Primrose, James, 146, 187

R

Rather, L.J., 181
Regius, Henricus, 142-3, 146-50, 154, 157, 160, 169, 177-8, 180, 187, 194
Reil, Johann Christoph, 206, 221
Riolan, Jean, 35, 57, 71, 83-4, 99, 101
Rosenfield, L., 186
Rothschuh, K.E., 30, 179, 181, 184, 187-9, 193, 218-9, 223
Rudbeck, Olaus, 184
Rüsche, F., 30, 110
Ruysch, Frederick, 189

S

Santorio, Santorio, 171
Sauvages, François Boissier de, 200, 218
Scherz, G., 184
Schimank, H., 177
Schütz, E., 230-1
Schultz, Carl Heinrich, 16, 198, 206-8, 221

Seneca, 95
Sennert, Daniel, 108
Servetus, Michael, 28, 44
Severinus, Petrus, 109
Singer, C., 15
Sloan, P.R., 16, 96, 182, 184-5, 187, 189
Spicker, S.F., 178
Stahl, Georg Ernst, 172, 200-1, 218, 226
Stannard, J., 97
Stannius, Hermann Friedrich, 217, 223
Starling, H., 227, 230
Stauffer, R.C., 178v
Steno (=Stennsen), Nils, 144, 164, 178, 184, 186, 193
Svammerdam, Jan, 186
Sylvius, Franciscus, 142-3, 150-4, 160, 173, 188, 192

T

Temkin, O., 30, 103-4, 178, 180
Thomas of Aquinas, 110, 178
Toellner, R., 17, 179
Treviranus, Johannes, 220

U

Underwood, E., 15
Unschuld, P.U., 9, 17

V

Vend, Georg Ernst, 222
Verbeek, T., 184, 187
Vesalius, Andreas, 28, 44, 181
Volkmann, Alfred Wilhelm, 214-7, 219, 223
Vorstius, A., 180, 186

W

Wale, Jan de, 142, 168-9, 194
Walther, Philipp Franz von, 197, 205, 212-3, 220
Ward, Seth, 185
Webster, C., 34, 93, 107, 114, 185
Weil, E., 185
Weinberg, K., 184
Westfall, R.S., 185
Whitteridge, G., 94-5, 97-9, 101-3, 107, 114, 185
Whytt, Robert, 218

Wilbrand, Johann Bernhard, 220
Willis, R., 94
Willis, Thomas, 142-3, 151, 156, 158, 160-3, 171, 174-5, 181, 192-3
Wolff, Caspar Friedrich, 197, 203, 219
Wolff, O., 230
Wren, Christopher, 158

Z

Zilsel, Edgar, 177
Zimmermann, Johann Georg, 200, 219